RAND NATIONAL DEFENSE RESEARCH INSTITUTE

T0146265

Assessing Bid Protests of U.S. Department of Defense Procurements

Identifying Issues, Trends, and Drivers

Mark V. Arena, Brian Persons, Irv Blickstein, Mary E. Chenoweth,
Gordon T. Lee, David Luckey, Abby Schendt

Prepared for the Office of the Secretary of Defense
Approved for public release; distribution unlimited

For more information on this publication, visit www.rand.org/t/RR2356

Library of Congress Cataloging-in-Publication Data is available for this publication.

ISBN: 978-1-9774-0005-5

Published by the RAND Corporation, Santa Monica, Calif.

© Copyright 2018 RAND Corporation

RAND® is a registered trademark.

Support RAND
Make a tax-deductible charitable contribution at
www.rand.org/giving/contribute

www.rand.org

Preface

This report documents the RAND Corporation's assessment of the prevalence and impact of bid protests on U.S. Department of Defense acquisitions. It is the product of a study on this issue that Congress called for in the National Defense Authorization Act for Fiscal Year 2017.[1] The findings are intended to inform Congress and U.S. defense leaders about the effectiveness of current procurement policies and processes to reduce bid protests. It assumes that the reader has some basic knowledge of the federal bid protest system and venues for filing protests.

RAND assembled and analyzed available data on bid protests and sought to address the study elements specified in Section 885 of the legislation. The analysis built on prior RAND research that assessed trends in U.S. Air Force bid protests, analyzed two high-profile bid protests (the Combat Search and Rescue Helicopter and Aerial Refueling Tanker Aircraft [KC-46] programs) for lessons learned, and recommended changes to Air Force acquisition tactics to counter bid protests in the future.[2] In addition, for the current study, the RAND research team reviewed and summarized studies and analyses conducted by the U.S. Government Accountability Office, the Congressional Research Service, and other organizations on the prevalence and impact of bid protests.

This report was delivered to Congress on December 21, 2017. It has since been professionally typeset and proofread.

This research was sponsored by the Under Secretary of Defense for Acquisition, Technology, and Logistics, Defense Procurement and Acquisition Policy, and conducted within the Acquisition and Technology Policy Center of the RAND National Defense Research Institute, a federally funded research and development center sponsored by the Office of the Secretary of Defense, the Joint Staff, the Unified Combatant Commands, the Navy, the Marine Corps, the defense agencies, and the Intelligence Community.

For more information on the Acquisition and Technology Policy Center, see www.rand.org/nsrd/ndri/centers/atp or contact the director (contact information is provided on the webpage).

---

[1] Public Law 114-328, National Defense Authorization Act for Fiscal Year 2017, December 23, 2016. Section 885 of the legislation requires a "comprehensive study on the prevalence and impact of bid protests on Department of Defense acquisitions."

[2] See Frank Camm, Mary E. Chenoweth, John C. Graser, Thomas Light, Mark A. Lorell, and Susan K. Woodward, *Government Accountability Office Bid Protests in Air Force Source Selections: Evidence and Options—Executive Summary*, Santa Monica, Calif.: RAND Corporation, MG-1077-AF, 2012, and Thomas Light, Frank Camm, Mary E. Chenoweth, Peter Anthony Lewis, and Rena Rudavsky, *Analysis of Government Accountability Office Bid Protests in Air Force Source Selections over the Past Two Decades*, Santa Monica, Calif.: RAND Corporation, TR-883-AF, 2012.

Contents

Figures

Tables

Summary

Bid protests have been a feature of the U.S. defense acquisition environment for decades. When interested parties that are providing goods or services to the U.S. Department of Defense (DoD) believe that the department has made an error in choosing the winning bid, they have the right to file protests questioning the outcome with the military services and defense agencies, the U.S. Government Accountability Office's (GAO's) Office of the General Counsel, or the U.S. Court of Federal Claims (COFC).[1]

Study Question

In recent years, DoD's bid protest process has come under increased scrutiny. Critics have argued that the department does not have a full understanding of the time and resources that it devotes to bid protests, the costs and schedule delays that it incurs throughout the process, or the incentives in the current process for companies to bid on defense business.[2] Critics have also argued that the current process may encourage frivolous protests and that DoD needs better information on the number, nature, and disposition of protests that it receives.

In response to this scrutiny, Congress—in the National Defense Authorization Act (NDAA) for Fiscal Year (FY) 2017—called for a "comprehensive study on the prevalence and impact of bid protests on DoD acquisitions."[3] The legislation further required the systematic collection and analysis of information on bid protests and their associated contracting outcomes and directed that the study take into account related input from DoD acquisition professionals.

RAND's Tasking

The RAND National Defense Research Institute was selected to conduct the study that Congress called for in the FY 2017 NDAA. The legislation requested an investigation of 14 elements

---

[1] For a definition of *interested party*, see U.S. Government Accountability Office, "Bid Protests at GAO: A Descriptive Guide," webpage, undated(b). We found several acronyms commonly used for the U.S. Court of Federal Claims. For consistency in this report, we use *COFC* as the shorthand reference.

[2] See for example, Charles S. Clark, "Conferees Will Determine Fate of Defense Bill Provision to Deter Frivolous Contractor Bid Protests," *Government Executive*, October 13, 2017; Christian Davenport, "Senate Proposes Measure to Curb Protests over Pentagon Contract Awards," *Washington Post*, October 8, 2017; and Steven J. Koprince, "Senate 2018 NDAA Re-Introduces Flawed GAO Bid Protest 'Reforms,'" *SmallGovCon*, July 28, 2017.

[3] See Section 885 of Public Law 114-328, National Defense Authorization Act for Fiscal Year 2017, December 23, 2016.

of the bid protest process to inform Congress and U.S. defense leaders about the effectiveness of procurement policies and processes that have been put in place to reduce and streamline protests. Of the 14 elements, we found sufficient information to address ten either fully or partially. These elements generally encompassed aspects of how the bid protest system affects or is perceived to affect DoD procurements, trends in bid protests, and differences in procurement characteristics. We were not able to address the four other elements—the effects of protests on procurements, the time and cost to the government to handle protests, the frequency with which a protester is awarded the disputed contract, and agency-level trends in protests—due to a lack of available data.

Study Approach

We pursued a two-pronged research approach to investigate the elements raised by Congress. One line of inquiry was qualitative and involved a literature review and a series of semi-structured discussions with experts in DoD's bid process.

Our literature review involved cataloging and identifying information related to the 14 elements of the bid protest process specified in the FY 2017 NDAA. We reviewed relevant open-source materials from a broad array of media outlets, research publications, and official public sources. In parallel, we held discussions with subject-matter experts from the U.S. government, industry, and the legal sector, as well as RAND experts who were knowledgeable about bid protests inside and outside of DoD.

Our second line of inquiry involved quantitatively examining DoD's current bid protest landscape. These efforts focused on collecting bid protest data and histories from GAO and COFC, the vast majority of which were based on their case dockets.[4] We also collected data from the military services and defense agencies; however, that information was not as robust as the GAO and COFC data.[5]

These data allowed us to examine and compare, for example, bid protest time trends, the number of protests per agency or per billion dollars of agency spending, the value of bids under protest, and the duration of bid protest proceedings. In addition, we were able to examine the number of protests filed by companies that had won prior bid solicitations, how and to what extent the prospect of bid protests affects the structure of DoD procurements, and the number and quality of pre- or post-award discussions and debriefings between DoD and bidders.

Summary Findings and Observations

Findings

Our qualitative analysis found substantial differences between how DoD and the private sector view these issues. In our discussions, DoD personnel expressed a general dissatisfaction

---

[4]   Derived from its docket system, GAO's record contained 21,186 actions related to protests, including actual protests, reconsideration requests, and requests for entitlements and cost. The record covered all government agencies under GAO's protest jurisdiction. The COFC data provided to the RAND team detailed 475 cases involving a DoD agency that were filed between January 2008 and May 2017. These records were compiled by the court clerk's office from its case docket system.

[5]   The data we collected tracked a variety of bid protest characteristics, including protests' primary agency, protester names, dates when cases were filed and closed, protests' disposition, the approximate value of the procurements at issue, and whether protesters were small or large businesses.

with the current bid protest system. They believed that contractors have an unfair advantage in the contracting process in that they are able to impede timely awards with bid protests. These personnel also stated that the protest rules encouraged this behavior by allowing protesters to make an excessive number of "weak" allegations, by permitting contractors too much time to protest, and by virtue of the amount of time it takes to resolve cases. In addition, there was a commonly held belief that a contractor is more likely to file a bid protest if it is an incumbent that has lost in a follow-on competition. The military services gave a variety of reasons for an incumbent filing a bid protest—ranging from structuring an orderly transition of its workforce to obtaining a follow-on bridge contract from the government that would provide additional revenue.

These DoD views were contrasted with views expressed in our discussions with representatives from private-sector companies, trade associations, and private law firms regarding the impact of bid protests on their corporate decisionmaking. Overall, the private sector views bid protests as a healthy component of a transparent acquisition process, because these protests hold the government accountable and provide information on how the contract award or source selection was made. If protests were not allowed or were curtailed, industry representatives argued, companies would likely make fewer bids. That said, a major private-sector concern was the quality of post-award debriefings. The worst debriefings were characterized as skimpy, adversarial, evasive, or failing to provide required reasonable responses to relevant questions. It became clear over the course of our study that too little information or debriefings that are evasive or adversarial may lead to a bid protest. The private sector also observed that the acquisition workforce was insufficiently staffed and could benefit from additional training. That workforce was cut massively in the 1990s and is still in the process of rebuilding.

Our quantitative analysis included a review of available data on bid protests filed with GAO and COFC.[6] We found a steady increase in the number of bid protest actions at GAO between FY 2008 and FY 2016; during that period, protest activity for both DoD and non-DoD agencies approximately doubled. Protest actions associated with DoD agencies accounted for roughly 60 percent of total protest actions over this period. We found a similar trend for protests filed with COFC. While there was a statistically significant increase in all protests at COFC over time, the upward trend in the number of DoD protests was not statistically significant.

The time trends are indifferent to changes in DoD spending and contracting. Overall DoD procurement data show that the number of contracts and contract spending declined from FY 2008 to FY 2016.[7] This trend runs counter to that for DoD bid protests. In Figure S.1, we show GAO data for the percentage of contracts protested and the number of contracts protested per billion dollars of DoD contract spending. The increases are statistically significant. Still, it is important to note that the overall percentage of contracts protested is very small—less than 0.3 percent. The trends are less clear at COFC, but the rates are an order of magnitude smaller (shown in Figure S.2). These small protest rates per contract imply that bid protests are exceedingly uncommon for DoD procurements.

What were the characteristics of those protests? Table S.1 shows a sample of the data we compiled and indicates that small businesses accounted for more than half of protest actions

---

[6]   Data on agency-level protests were not available.

[7]   Based on data from the Federal Procurement Data System–Next Generation (FPDS-NG).

Figure S.1
Percentage of DoD Procurements Protested and Protests per Billion Dollars at GAO

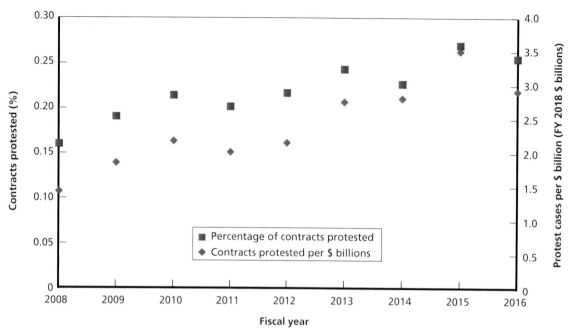

SOURCE: RAND analysis of GAO and FPDS-NG data.
RAND *RR2356-S.1*

Figure S.2
Percentage of DoD Procurements Protested and Protests per Billion Dollars at COFC

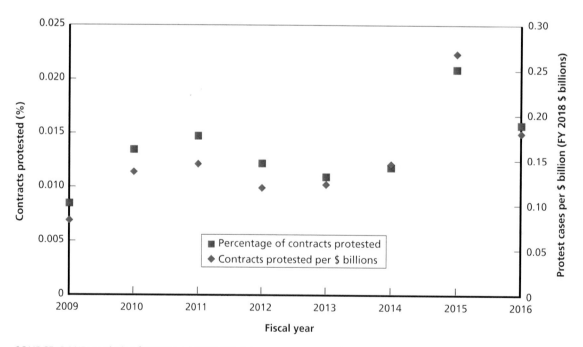

SOURCE: RAND analysis of COFC and FPDS-NG data.
NOTE: Complete data for FY 2008 were not available.
RAND *RR2356-S.2*

at GAO and COFC.[8] The implication is that any changes or improvements to the bid protest system need to account for small businesses. Improvements aimed only at larger firms would miss the majority of DoD bid protest actions at GAO and COFC.

Table S.1 also reveals that 4–8 percent of protest actions were associated with procurements valued by the protesters at under $0.1 million—with some at the micro-purchase level.[9]

Observations

It should be noted that the overall percentage of DoD contracts protested was very small—less than 0.3 percent. This small percentage implies that bid protests are exceedingly uncommon for DoD procurements. However, it also should be noted that overall protest activity is increasing at both COFC and GAO and that small-business protests represent the majority of protests at both venues.

An overarching conclusion from our research (both our discussions and our statistical analysis) is that policymakers should avoid drawing overall conclusions or assumptions about

Table S.1
DoD Bid Protest Characteristics

Characteristic	GAO (all actions)	COFC (case data)
Observations	11,459	475
Time frame	FY 2008–FY 2016	CY 2008–CY 2016
From small businesses[a]	53%	58%
Value under $0.1 million[a]	7.9%	3.5%
Task-order protests	10.6%	NA[b]
Sustained rate[c]	2.6%	9%
Effectiveness rate	41%	NA[d]
Average time to close (days)	41	133

SOURCE: RAND analysis of GAO and COFC data.

NOTES: We defined the sustained rate as the number of actions in which GAO issued a decision in favor of the protester relative to all protest actions. The effectiveness rate is the number of protest actions that were either sustained or assigned corrective action relative to all protest actions. CY = calendar year.

[a] Self-reported by the protester.

[b] COFC generally does not have jurisdiction over task-order protests.

[c] Note that GAO measures this rate relative to merit cases and not total protest actions due to the relatively high overall rate of agency corrective actions. Using GAO's method, the value is 12 percent.

[d] Data on corrective actions at COFC were not available.

---

[8] A caveat here is that whether the protester is a small business is self-reported by the protester. Therefore, it is unclear whether all the firms identified as small businesses actually qualify under the formal definition of the Small Business Administration. Nonetheless, the majority of protesting firms in our data self-identified as small businesses.

[9] The value of "less than $100,000" was the lowest value category provided by GAO (aside from micro-purchases). Micro-purchases amounted to $3,500 or less over the time frame of the protest data. See question 2 at Office of the Secretary of Defense, Defense Procurement and Acquisition Policy, "Government Purchase Card (GPC) Frequently Asked Questions (FAQs)," webpage, last updated October 17, 2017.

trends from one case when it comes to the efficacy of the bid protest system. Policymakers should be aware that the perspectives of the bid protest system held by DoD personnel and by the private sector vary greatly and that there is a lack of trust on each side. Our research pointed to several other, more specific observations and recommendations.[10]

The data we reviewed pointed to several GAO-specific observations:

- The stability of the bid protest effectiveness rate over time—despite the increase in protest numbers—suggests that firms are not likely to protest without merit.
- Small-business protests are less likely to be effective and more likely to be dismissed for legal insufficiency.
- Protest filing peaks at the end of the fiscal year.
- Task-order protests have a slightly higher effectiveness rate than other types of protests.
- There are measurable differences between the services and defense agencies, but, overall, DoD (services and agencies) has a slightly lower effectiveness rate than non-DoD agencies.
- The largest DoD contractors have slightly higher sustained and effectiveness rates, but these differences are diminishing with time.
- Cases in which legal counsel is required (i.e., a protective order was issued by GAO) have higher effectiveness and sustained rates.[11]
- DoD uses stay overrides infrequently.[12]
- The number of protesters and protest actions tends to grow with a contract's value.

The data we reviewed also pointed to several COFC-specific observations:

- The sustained rate at COFC is declining with time as the number of cases increases. These trends suggest that firms may be more willing to file protests with COFC.
- There are no differences in sustained rates between DoD components and agencies or between small and larger businesses.
- The appeals rate is declining over time.
- Data and discussions suggest that the number of COFC cases that previously appeared at GAO may be increasing, but this potential trend needs further research.

Finally, the analyses presented in Chapters Four and Five of this report point to several observations common to both GAO and COFC:

- While our statistical modeling indicates differences between types of cases (categorization of cases), it is not possible to predict the outcome of any case based on its general characteristics. Each case is different and its details affect the outcome.
- The overall level (numbers) of bid protest activity (DoD and non-DoD) has been increasing at both GAO and COFC since 2008.

---

[10] All differences and trends are statistically significant unless otherwise noted.

[11] When the administrative record contains proprietary, confidential, or source selection–sensitive information, GAO may issue a protective order (generally at the request of the protester) to allow the protester's attorneys access to the information.

[12] A stay override occurs when an agency overrides the automatic hold of execution (award or performance) during a protest at GAO.

- Bid protests by plaintiffs that identify themselves as small businesses represent the majority of protests at both venues.
- At both venues, in a nontrivial number of cases (approximately 4–8 percent), the contract value is less than $0.1 million (then-year dollars, as reported by the protester).
- There are differences between the services and DoD agencies in terms of the number of cases filed. Specifically, the Army has the highest number of cases, year-on-year, at both venues. This is partly explained by its share of contract expenditures.
- Trends differ between GAO and COFC, suggesting that any changes to the protest system should be tailored to the venue. For example, COFC's sustained rate is declining, whereas at GAO it is holding steady (and potentially increasing).

Recommendations

From the observations derived from our qualitative and quantitative analyses, we generated the following recommendations for policymakers and DoD leadership:

- **Enhance the quality of post-award debriefings.** The Army and Air Force have initiatives to improve the quality of the debriefings, which might serve as models. Section 818 of the FY 2018 NDAA has provisions for improving debriefings as well.
- **Be careful in considering any potential reduction to GAO's decision timeline.** While 70 percent of cases at GAO are resolved in less than 60 days, it may be challenging to shorten the GAO decision timeline for all cases given that (1) protests are more frequently filed at the end of the fiscal year and (2) complex cases that go to decision usually take 90–100 days.
- **Be careful in considering any restrictions on task-order bid protests at GAO.** Task-order protests have a slightly higher effectiveness rate than the rest of the protest population. This higher rate suggests that there may be more challenges with these awards and that task-order protests fill an important role in improving the fairness of DoD procurements.
- **Consider implementing an expedited process for adjudicating bid protests of procurement contracts with values under $0.1 million.** One possible option is a process analogous to how traffic tickets are adjudicated in traffic court or how cases are adjudicated in small-claims court. A different approach would likely be needed for each venue. For example, COFC could "rule from the bench" on such smaller-value protests and not be required to generate written decisions. (This would limit the protester's ability to appeal, however.) Another option is to require alternative dispute resolution for such small-value protests at GAO. Some discussion with each venue would be necessary to develop the most appropriate approach. Another but perhaps less desirable approach from a fairness perspective would be to restrict such low-value procurement protests to the agency level. Our recommendation is to come up with a quick way to resolve these cases commensurate with their value while preserving the right to an independent protest.
- **Consider approaches to reduce and improve protests from small businesses,** such as improving debriefings, requiring protests to be filed by legal counsel, or providing legal assistance in filing.

- **Consider collecting additional data and making other changes to bid protest records to facilitate future research and decisionmaking.** Some examples include tracking cases that appear at COFC with a prior history at GAO, recording companies' DUNS numbers, tracking corrective action at COFC, collecting and summarizing the reasons for corrective action, and generating annual reports of agency-level protest activity.

These recommendations are intended to inform future changes to the bid protest system. There is likely value in using the same or similar approaches across other departments and agencies of the U.S. government. In implementing these recommendations, there should be some consideration of costs and benefits, as some changes will require additional time or resources to implement.

Acknowledgments

Unfortunately, we cannot individually acknowledge all those who gave their time and wisdom to assist us in this research. We thank the multiple individuals from the private sector and affiliated associations and the many practicing government contracting attorneys who offered their time and insight into the bid protest system. Likewise, we thank the various acquisition representatives from the services and the Defense Logistics Agency for sharing their views and data.

Without the help of GAO and COFC, much of this analysis would not have been possible. We would like to specifically thank Edward Goldstein, Kenneth Patton, and Ralph White, Jr., at GAO for providing an extensive amount of bid protest data, as well as answering our innumerable questions. Likewise, we are grateful to many individuals at COFC—particularly Chief Judge Susan Braden—for their help, information, and insight. We also thank Jacob Wilson and Summer Maynard for their assistance in collecting and organizing bid protest data for the court.

Moshe Schwartz of the Congressional Research Service provided very helpful advice throughout our research effort. Similarly, we thank Douglas Buettner in the Office of the Secretary of Defense for his help and reviews during this study. Our study monitor, Gregory Snyder in the Office of the Secretary of Defense, Defense Procurement and Acquisition Policy, deserves special attention for facilitating our interactions with government personnel and helping the research run smoothly.

We also thank our RAND reviewers, Thomas Light and Philip Anton, for their many helpful suggestions and clarifications. The report is vastly improved as a result. Finally, we thank RAND editor Lauren Skrabala, who significantly advanced the readability of this report.

Abbreviations

ADR	Alternative Dispute Resolution
CICA	Competition in Contracting Act of 1984
COFC	U.S. Court of Federal Claims
CY	calendar year
DLA	Defense Logistics Agency
DoD	U.S. Department of Defense
FAR	Federal Acquisition Regulation
FPDS-NG	Federal Procurement Data System–Next Generation
FY	fiscal year
GAO	U.S. Government Accountability Office (since July 7, 2004); U.S. General Accounting Office (prior to July 7, 2004)
LPTA	lowest price technically acceptable
NDAA	National Defense Authorization Act
OCO	overseas contingency operations
RFP	request for proposal
RFQ	request for quote

CHAPTER ONE

Introduction

This report assesses the prevalence and impact of bid protests on U.S. Department of Defense (DoD) acquisitions of systems and services. Its goal is to inform Congress and U.S. defense leaders about the effectiveness of procurement policies and processes that aim to reduce protests.

Bid protests are a long-standing feature of the U.S. defense acquisition environment. When a bidder (also referred to as an offeror) in a source selection believes that DoD has made an error that is large enough to change the outcome of the selection, that bidder has the right to file a protest with the U.S. Government Accountability Office's (GAO's) Office of the General Counsel or with the U.S. Court of Federal Claims (COFC).[1] Actions filed with GAO trigger a review by that office; if GAO agrees that a significant error has occurred and has the potential to change the source selection outcome, it can suggest a course of remediation to DoD. Protests filed with COFC initiate the creation of a case. The outcome of the case is legally binding and the appropriate course of action is issued as a decision by the court.

The bid protest process came under greater congressional scrutiny in 2017. In Section 885 of the National Defense Authorization Act (NDAA) for Fiscal Year (FY) 2017, Congress required a "comprehensive study on the prevalence and impact of bid protests on Department of Defense acquisitions."[2] The act called for the systematic collection and analysis of data on bid protests, their associated outcomes, and their impact on DoD procurements.

This report fulfills that NDAA requirement. It assembles and analyzes available data on bid protests and seeks to address 14 study elements laid out in Section 885 of the FY 2017 NDAA. When available data did not support the analysis called for in the legislation, we recommend changes to current efforts to track and monitor bid protest activity to generate this information in the future and meet the requirements for this type of analysis.

How RAND Conducted the Study

To address the questions raised by Congress relative to bid protests of DoD procurements, we assembled a research team that pursued a two-pronged analytical approach:

- a literature review and discussions with dozens of experts who were knowledgeable about DoD's bid process to understand the effects and perceptions of the bid protest system on decisionmaking

---

[1] We found several acronyms commonly used for the U.S. Court of Federal Claims. For consistency's sake in this report, we use *COFC* as the shorthand reference.

[2] Public Law 114-328, National Defense Authorization Act for Fiscal Year 2017, December 23, 2016.

- a quantitative evaluation of bid protest trends that drew on primary-source data from GAO and COFC.

Our literature review involved cataloging and identifying information related to the 14 statutorily mandated issue areas in the FY 2017 NDAA. Our open sources included materials from a broad array of media outlets, research publications, and official public sources. We focused our examination on information relevant to the 14 elements under consideration. The subject-matter experts with whom we held discussions included selected U.S. government personnel, public-sector representatives (from both industry and the legal profession), and RAND experts with experience studying bid protests in DoD and elsewhere in the federal government. From our literature review and discussions, we developed findings and recommendations for DoD that are generally applicable to other federal departments and agencies.

Our quantitative analysis involved collecting bid protest data and histories from GAO and COFC based on their case dockets. We also attempted to collect data from the military services, but there was great disparity in the type of protest information that each service collects centrally. The data from each of the protest venues—GAO and COFC—were generally more comprehensive. We also requested data on agency-level protests, but none of the services tracked these protests centrally.

Other aspects of the NDAA requirement that we addressed were the extent and manner in which the bid protest system affects or is perceived to affect procurement in such a way as to avoid protests rather than improve acquisition; the quality and number of pre-proposal discussions, discussions of proposals, or post-award debriefings; decisions to use lowest-price-technically-acceptable (LPTA) procurement methods, to make multiple awards or encourage teaming, to use sole source-award methods, and to exercise options on existing contracts; and the ability to meet an operational or mission need or to address important requirements.

Bid Protest Issue Areas That Congress Directed Be Studied

Section 885 of the FY 2017 NDAA called for a study to address 14 elements surrounding bid protests. However, for some of those areas, data and information were not available. For example, the U.S. military services and other agencies do not collect data on costs associated with addressing bid protests. Most DoD agencies are mission-funded and do not have activity-based accounting systems to track protest activity at the level of fidelity required by Section 885 of the FY 2017 NDAA.[3] In conversations with DoD personnel, we learned that these organizations do not track the costs associated with filing a bid protest; if they did, they were reluctant to provide that information.

Table 1.1 lists the 14 study elements and notes whether we were able to obtain necessary data and information.

---

[3]   A typical response from the services and agencies we contacted on this matter was, "If we get a bid protest, we just deal with it as part of executing the overall mission of the organization." The same held true for representatives from the private sector.

Table 1.1
Fourteen Study Requirements

Element Number	Element Description in the FY 2017 NDAA	Data/Information Available
1	For employees of the Department [of Defense], including contracting officers, program executive officers, and program managers, the extent and manner in which the bid protest system affects or is perceived to affect: (A) the development of a procurement to avoid protests rather than improve the acquisition; (B) the quality or quantity of pre-proposal discussions, discussions of proposals, or post-award debriefings; (C) the decision to use lowest price technically acceptable procurement methods; (D) the decision to make multiple awards or encourage teaming; (E) the ability to meet an operational or mission need or address important requirements; (F) the decision to use sole source award methods; and (G) the decision to exercise options on existing contracts.	Yes. Data and information were obtained for all subelements through discussions with relevant DoD employees.
2	With respect to a company bidding on contracts or task delivery orders, the extent and manner in which the bid protest system affects or is perceived to affect— (A) the decision to offer a bid or proposal on single award or multiple award contracts when the company is the incumbent contractor; (B) the decision to offer a bid or proposal on single award or multiple award contracts when the company is not the incumbent contractor; (C) the ability to engage in pre-proposal discussions, discussions of proposals, or post-award debriefings; (D) the decision to participate in a team or joint venture; and (E) the decision to file a protest with the agency concerned, the Government Accountability Office, or the Court of Federal Claims.	Yes. Data and information were obtained for all subelements from discussions with trade association representatives (e.g., National Defense Industrial Association, Aerospace Industries Association, Professional Services Council) on behalf of their member companies.
3	A description of trends in the number of bid protests filed with agencies, the Government Accountability Office, and Federal courts, the effectiveness of each forum for contracts and task or delivery orders, and the rate of such bid protests compared to contract obligations and the number of contracts.	Partial. Agency-level protest data were not available.
4	An analysis of bid protests filed by incumbent contractors, including— (A) the rate at which such protesters are awarded bridge contracts or contract extensions over the period that the protest remains unresolved; and (B) an assessment of the cost and schedule impact of successful and unsuccessful bid protests filed by incumbent contractors on contracts for services with a value in excess of $100,000,000.	Partial. We were able to examine incumbent and schedule implications of protests by supplementing GAO data with additional analysis of data from Deltek's GovWin database. Cost impact information was not available.
5	A comparison of the number of protests, the values of contested orders or contracts, and the outcome of protests for— (A) awards of contracts compared to awards of task or delivery orders; (B) contracts or orders primarily for products, compared to contracts or orders primarily for services; (C) protests filed pre-award to challenge the solicitation compared to those filed post-award; (D) contracts or awards with single protestors compared to multiple protestors; and (E) contracts with single awards compared to multiple award contracts.	Partial. GAO and COFC data analysis for all subelements except E.
6	An analysis of the number and disposition of protests filed with the contracting agency.	No. Data were not available.

Table 1.1—Continued

Element Number	Element Description in the FY 2017 NDAA	Data/Information Available
7	A description of trends in the number of bid protests filed as a percentage of contracts and as a percentage of task or delivery orders awarded during the same period of time, overall and set forth separately by the value of the contract or order as follows: (A) Contracts valued in excess of $3,000,000,000. (B) Contracts valued between $500,000,000 and $3,000,000,000. (C) Contracts valued between $50,000,000 and $500,000,000. (D) Contracts valued between $10,000,000 and $50,000,000. (E) Contracts valued under $10,000,000.	Yes. GAO and COFC data analysis. However, GAO does not track contract values in these ranges. We used GAO-reported ranges instead.
8	An assessment of the cost and schedule impact of successful/unsuccessful bid protests filed on contracts valued in excess of $3,000,000,000.	Partial. Cost data were not available. We were able to examine schedule impact based on a subset of GAO data but not for the precise value range requested.
9	An analysis of how often protestors are awarded the contract that was the subject of the bid protest.	No. We were not able to align protest data with the successful awardee. We were able to determine only whether the protest was successful.
10	Summary of the results of protests in which the contracting agencies took unilateral corrective action, including— (A) at what point in the bid protest process the agency agreed to take corrective action; (B) the average time for remedial action to be completed; and (C) a determination regarding— (i) whether or to what extent the decision to take the corrective action was a result of a determination by the agency that there had been a probable violation of law or regulation; or (ii) whether or to what extent such corrective action was a result of some other factor.	Partial. We were able to address subelement A for GAO data and subelement B for a limited set of GAO data. Most agencies were not able to provide data for subelement C, or they viewed this information as privileged.
11	A description of the time it takes agencies to implement corrective actions after a ruling or decision, and the percentage of those corrective actions that are subsequently protested, including the outcome of any subsequent protest.	No. Data were not available.
12	An analysis of those contracts with respect to which a company files a protest (referred to as the "initial protest") and later files another protest (referred to as the "subsequent protest"), analyzed by the forum of the initial protest and the subsequent protest, including any difference in the outcome, between the forums.	Partial. We were able to examine subsequent protest actions at GAO. We were not able to fully track protests between venues at the case level.
13	An analysis of the effect of the quantity and quality of debriefings on the frequency of bid protests.	Yes. Data were obtained through discussions.
14	An analysis of the time spent at each phase of the procurement process attempting to prevent a protest, addressing a protest, or taking corrective action in response to a protest, including the efficacy of any actions attempted to prevent occurrence of a protest.	No. Data were not available.

SOURCE: Pub. L. 114-328, 2016, Section 885 (columns 1–2), and RAND research (column 3).

Organization of This Report

In Chapter Two, we detail our review of the literature on the bid protest process and its history. In Chapter Three, we discuss stakeholder perspectives of the bid protest process, which we derived from our discussions with subject-matter experts from the military services, government agencies, and the private sector. That is followed in Chapter Four by a discussion of trends and features of bid protests at GAO. Chapter Five describes similar issues for protests at COFC. Chapter Six explores a subset of the protest data to address some of the requirements outlined in Section 885 of the FY 2017 NDAA. Chapter Seven provides observations and recommendations. Appendix A offers additional detail on our analysis of GAO and COFC data. Appendix B describes our review of contract activity using data from the Federal Procurement Data System–Next Generation (FPDS-NG). Appendix C provides a list of the questions that guided our discussions with DoD personnel.

CHAPTER TWO

Bid Protest Definition, Brief Historical Overview, and Related Research

This chapter presents findings from our literature review to define bid protests, trace the federal government's history with the process, and review agencies' goals and the driving theory behind current practices. These explorations of the definition of bid protests, their history, and approaches to fielding them set the stage for the assessments and evaluations that follow.

Defining Bid Protests

Bid protests are challenges to the terms and conditions of a solicitation, an award decision, or a decision to cancel a solicitation or award. Federal Acquisition Regulation (FAR) 33.101 specifically defines a protest as

> a written objection by an interested party to any of the following:
>
> (1) A solicitation or other request by an agency for offers for a contract for the procurement of property or services.
> (2) The cancellation of the solicitation or another request.
> (3) An award or proposed award of the contract.
> (4) A termination or cancellation of an award of the contract, if the written objection contains an allegation that the termination or cancellation is based in whole or in part on improprieties concerning the award of the contract.[1]

A protest may be filed only by an interested party, generally a representative of a company whose direct economic interest has been or would be affected by the issues that motivated the protest. Proposed subcontractors and other companies that did not submit proposals before the due date are not interested parties. Typical issues that interested parties raise fall into two categories: *pre-award protests* and *post-award protests*.

Pre-award protests generally raise issues about the terms and conditions of the solicitation. They may challenge an interpretation of specific language in the solicitation or decisions to restrict competition. *Post-award protests* typically challenge the evaluation process by arguing that the soliciting agency failed to follow evaluation criteria; that the evaluation violated procurement law, regulations, or policies; or that the award was arbitrary and capricious or exhibited an abuse of discretion. Another type of post-award protest is a challenge to the size

---

[1] Federal Acquisition Regulation, Subpart 33.1, "Protests," May 29, 2014, section 33.101, "Definitions."

or eligibility of an awardee receiving a contract set aside for small businesses or other special classes of business.[2]

Agency-Level Bid Protests

Bid protests may be filed with any of several forums or venues.[3] Many agencies (DoD and non-DoD) have a process for filing an informal protest with a designated contracting officer or other official. These are designated as *agency-level bid protests.*

Formal bid protests may also be filed with GAO or COFC for all procurements subject to the FAR. The majority of DoD procurements are subject to the FAR, and any formal bid protests are accordingly filed at GAO or COFC.[4]

DoD allows a contractor to file an agency-level protest to resolve an issue at the lowest level possible. This concept has its origins in Title 48, Section 33.103, of the Code of Federal Regulations and Executive Order 12979, issued in the mid-1990s, which directed agencies to prescribe administrative procedures to resolve bid protests as an alternative to bid protests filed outside the agency.[5] The executive order further emphasized that agency-level protests should—to the maximum extent practicable—provide an inexpensive, informal, procedurally simple, and expeditious resolution of bid protests, including, when appropriate and as permitted by law, the use of alternative dispute resolution techniques, third-party neutrals, and other agency personnel.[6]

FAR 33.103, titled "Protests to the Agency," implements Executive Order 12979 and includes many details regarding filing deadlines and required information. An agency-level bid protest is intended to be an efficient, informal way for a contractor and contracting officer to clarify and amicably resolve their issues. It is supposed to take a fraction of the time to resolve a protest in this manner—with an objective (that is not binding) of less than 35 days versus up to 100 days at GAO or even longer at COFC. An unsuccessful offeror is usually reluctant to file an agency-level bid protest because the relationship with the contracting officer may already be strained and the offeror may feel that the contracting officer and agency will not be able to render an impartial, objective decision.[7] Also, the time to resolve an agency-level protest may affect the timeliness restrictions for a stay under the Competition in Contracting Act of 1984 (CICA) if the protest is subsequently filed with GAO.[8]

---

[2] For a more in-depth discussion of bid protests, see Daniel I. Gordon, "Bid Protests: The Costs Are Real, but the Benefits Outweigh Them," *Public Contract Law Journal,* Vol. 42, No. 3, Spring 2013.

[3] For a review of bid protest venues and their differences, see Michael J. Schaengold, Michael Guiffre, and Elizabeth M. Gill, "Choice of Forum for Federal Government Contract Bid Protests," *Federal Circuit Bar Journal,* Vol. 18, No. 243, 2009. Also see William E. Kovacic, "Procurement Reform and the Choice of Forum in Bid Protest Disputes," *Administrative Law Journal of American University,* Vol. 9, 1995.

[4] Interagency Alternative Dispute Resolution Working Group, "Electronic Guide to Federal Procurement ADR 2d," webpage, undated(a).

[5] Code of Federal Regulations, Title 48, "Federal Acquisition Regulations System," Section 33.103, "Protests to the Agency"; Executive Order 12979, *Agency Procurement Protests,* October 25, 1995.

[6] Executive Order 12979, 1995.

[7] For a more detailed discussion, see Daniel I. Gordon, "Constructing a Bid Protest Process: Choices That Every Procurement Challenge System Must Make," *Public Contract Law Journal,* Vol. 35, No. 3, Spring 2006.

[8] Interagency Alternative Dispute Resolution Working Group, undated(a).

Bid Protests Filed with GAO or COFC

When an offeror believes that DoD has made an error that is large enough to change the outcome of the source selection and feels that the department cannot render an impartial and objective decision through an agency-level bid protest, the offeror can file a protest with GAO's Office of the General Counsel or with COFC.

In 1985, GAO drafted detailed regulations governing bid protests for federal procurements. These regulations were promulgated to implement CICA. The regulations have been continuously revised to reflect statutory and other changes. In January 2003, the regulations were revised to conform to current practice and otherwise improve the efficiency and efficacy of the bid protest process at GAO. A bid protest filed with GAO automatically stays the award of the contract until the case is closed. However, the contracting agency can override the stay for urgent and compelling reasons, such as a national emergency or if immediate procurement is in the best interest of the United States. However, it does so under the risk that GAO will uphold the bid protest. Agencies rarely proceed with contracts when there is an active bid protest (as will be shown in the following chapters). Following its review, GAO can suggest a course of remediation to DoD, if it agrees that a significant error has occurred and that the error has the potential to change the source selection outcome.[9]

An offeror may also elect to file a bid protest with COFC either initially or after it has filed a bid protest with GAO. (This can be done before or after a case is resolved at GAO; however, if it is done before, the case is automatically dismissed at GAO.) Because GAO is not a court but, rather, a legislative branch agency, COFC is not bound by GAO's recommendations. A bid protest at COFC does not automatically stay the award of the contract as in the case of protests filed with GAO. An injunction to stay the contract award, however, can be filed with COFC. Unlike GAO filings, in which offerors can represent themselves, at COFC, an offeror must be represented by authorized legal counsel. Moreover, the administrative record that is required at COFC is more comprehensive than that typically required by GAO. Consequently, costs are presumably greater for offerors that file bid protests with COFC. Decisions promulgated by COFC are considered final and can be appealed only to the U.S. Court of Appeals or, in extremely rare cases, the U.S. Supreme Court.

The timeline that each of the formal protest venues follows is different. We go into more detail on the timeline for each venue in Chapters Four and Five. Once a protest is filed at GAO, the protest details are reviewed for legal sufficiency, timeliness, jurisdiction, and so forth. Protests that do not meet these various requirements are typically dismissed quickly. For cases that proceed, the corresponding agency has 30 days to file a report responding to the protest and to provide relevant details from the administrative record. The agency may take corrective action at any time, which results in the case being dismissed or withdrawn. When corrective action occurs, it is typically before the administrative record is due. After the report is filed, the protester must respond to the report. If the case proceeds, GAO may hold hearings or other informational meetings to inform its decision. Cases that go to decision (also known as merit cases) result in GAO sustaining or denying the protest. A decision by GAO must be complete within 100 days of the protest being filed (barring unusual circumstances, such as a government shutdown).

---

[9] U.S. Government Accountability Office, *Bid Protests at GAO: A Descriptive Guide*, 9th ed., Washington, D.C., GAO-09-471SP, 2009a.

The protest process at COFC is "motion-oriented" and begins when a protester files a complaint. The subsequent timing of the case revolves, in part, around the filing of the administrative record by the defending government agency. Shortly after the filing, the assigned judge will hold a scheduling conference to set the timing for the protest case and to determine the status of the procurement. The court also determines early in the case whether the government will voluntarily maintain the status quo on the procurement or whether it will possibly grant a restraining order if warranted and requested by the protester. Once the administrative record is filed, the parties respond and motions are filed by all parties for various outcomes (e.g., dismissal of the case, discovery, judgment on the administrative record). After all motions and responses have been filed, the court holds oral arguments in which parties present their views. Once oral arguments are complete, the court rules on the case in a written decision.

History of Bid Protests

The Tucker Act of 1887 was one of the earliest legislative acts regarding U.S. government contracting. Through the act, the U.S. government waived its sovereign immunity in certain lawsuits. It allowed for lawsuits in cases of "express or implied" contracts in which the U.S. government was a party. Prior to the Tucker Act, the U.S. government could not be sued.

The framework for the bid protest system began to take shape when GAO was established as the U.S. General Accounting Office in the Office of the Comptroller General through the Budget and Accounting Act of 1921. Prior to this, bid protests were directed through the judicial branch, often with little success for protesters.[10] It should be noted, however, that few if any protests were successful in this new venue before the 1950s. The reason appears to be that the government was not required or expected to meet additional requirements; rather, the government was free to act similarly to a private company in its distribution of contracts. Through the 1950s, GAO acted as the only nonjudicial venue for protests outside of the contracting agency.

The Walsh-Healey Public Contracts Act of 1936 was passed as a part of the New Deal and dealt with labor rights for workers on government contracts. Its goal was to improve general working conditions, and it applied to goods-based government contracts exceeding $10,000. The act established overtime pay, minimum wage, and safety standards and prohibited labor from convicts and persons younger than 16. It was the follow-up to Executive Order 6246, which required government agencies to work under fair competition requirements. The act was largely considered a signal of "good faith" that the government would honor and enforce its own labor laws in the award of contracts.

The Administrative Procedure Act of 1946 set a standard for judicial review. The legislation "[g]overns the process by which federal agencies develop and issue regulations."[11] Additionally, it "[p]rovides standards for judicial review if a person has been adversely affected or aggrieved by an agency action."[12] The act was especially important in that it extended the powers of judicial review beyond the assessment of financial damages or intent to commit fraud.

---

[10] The legal side of this arrangement is discussed in greater detail later in this chapter.

[11] U.S. Environmental Protection Agency, "Summary of the Administrative Procedure Act: 5 USC §551 et seq. (1946)," webpage, last updated December 30, 2016.

[12] U.S. Environmental Protection Agency, 2016.

The Wunderlich Act of 1954 arose out of the 1951 Supreme Court case *United States v. Wunderlich*. This legislation provided that, should a contractor appeal an administrative decision to a court, any administrative determinations under a disputes clause "shall be final and conclusive."[13] A later Supreme Court ruling, *United States v. Carlo Bianchi & Co.*, ruled that, according to the Wunderlich Act, the court could not make a determination of the facts and must confine its review to the administrative record.

Some three decades after the Wunderlich Act, Congress passed the Federal Courts Improvement Act of 1982, which established the U.S. Court of Appeals for the Federal Circuit and the U.S. Claims Court, which would later become COFC.

Shortly thereafter, when Congress passed the Competition in Contracting Act of 1984, a journal article clarified the new view of the role of government in contracting:

> Because many businesses make Government contracting their sole or principal source of income, the award of a contract may be vitally important if follow-on work is obtained. Similarly, a growing dependence on Government contracting for its source of income may cause a business to seek a contract in order to keep its plant facilities operating or its personnel employed. The Government contract, therefore, has for many businessmen a value that must measure by more than profit that flows directly to a company as the result of performing a specific contract.[14]

This legislation ultimately expanded and augmented GAO's power to resolve bid protests. It also established—for a trial period—the General Services Board of Contract Appeals for protests involving data processing equipment and telecommunications. The Paperwork Reduction Act of 1986 made the board a permanent bid protest jurisdiction.

Goals and Theory Behind Bid Protests

The government's involvement in the bid protest system has evolved to meet the perceived and real role that its contracting arm plays in the economy. As a result, several theories have arisen to support the complex bid protest system in its current form.[15]

The first theory, which emerges from the discussion relayed in the prior section on the history of bid protests, is that the government is a powerful entity in the economy and, as such, has a moral duty to maintain fairness in how it awards large contracts. Bid protests attempt to "accomplish nonefficiency goals that ordinarily are of little concern to private firms."[16] Taxpayers typically want their government to deal fairly when it distributes money, judgments, and other "services" paid for by taxpayer money. Bid protests aim to ensure that government purchasing agents deal "fairly" with prospective suppliers. Public funds come with expectations of transparency and equity in how they are distributed. While it is not expected that a private

---

[13] U.S. Code, Title 41, Public Contracts, Section 321, Limitation on Pleading Contract Provisions Related to Finality; Standards of Review.

[14] George M. Coburn, "The New Bid Protest Remedies Created by the Competition in Contracting Act of 1984," *Journal of Contract Management*, Vol. 19, Summer 1985.

[15] Gordon, 2006.

[16] Kovacic, 1995, p. 468.

firm will always choose the best or cheapest option, as it may have preexisting partnerships that specifically fit its business model, taxpayers expect that the U.S. government will provide all offerors with an "equivalent" chance.

This idea that the government's use of funds must be held to a higher standard goes beyond the idea of fairness. Taxpayers are also concerned with integrity, and, as such, the federal government should ensure that the process and methods by which it allocates funds exhibit the highest possible degree of integrity. U.S. agencies are held to a different set of standards than their private counterparts, simply because they are using government funds. By that measure, protests and control measures are in place to "deter and punish ineptitude, sloth, or corruption of public purchasing officials."[17]

The second theory underlying the current bid protest system is that officials allocate contracts with public funds and do not experience the same incentives that they might with their own agencies' money. Private firms have their own method for compensating for the weaker incentives of their agents (i.e., compensation schemes and monitoring). But in the public sector, bid protests are designed to compensate for such weaker incentives.

From the perspective of potential offerors, a third theory holds that the protest system acts as a signaling mechanism to potential private partners. Government contracting bears unique risks that are absent in private or commercial contracts. For example, before a contract is awarded, companies must make a significant investment to meet the unique needs of the government.[18] The existence of a system for private companies to lodge a complaint or to protest a contract decision signals that the government is a suitable partner; such a system shows that if a decision is perceived to be unfair, protesters can appeal the decision and have another party review it.

Review of Quantitative Research on Bid Protest Activity

There is a significant amount of scholarly research on bid protest activity. So much so that we cannot cite it all.[19] In this section, we highlight recent research on bid protests that included some quantitative analysis and that was less focused on case studies or anecdote. This overview provides context for our analyses in Chapters Four, Five, and Six and focuses on addressing the questions posed in Section 885 of the FY 2017 NDAA.

In terms of GAO bid protests, the office itself publishes an annual report summarizing bid protest outcomes and trends (outcome statistics, filing activity, and reasons that cases were sustained).[20] These annual reports serve as source material for much of the other research on

---

[17] Kovacic, 1995, p. 469.

[18] These investments may be in how firms bill to a contract. It is often the case that commercial contracts pay a certain amount of money for services and do not dictate how that money is spent as long as the services are provided as specified. Government contracts, however, require that all funds spent on a contract be managed and tracked so that auditors can review the records and ensure that money is being spent appropriately. On the other side of the second bid protest theory, "If government purchasing officials in fact have weaker incentives than private buyers to make efficient procurement choices, prospective offerors also may perceive a greater risk that purchasing decisions will be made arbitrarily" (Kovacic, 1995, p. 468).

[19] A more complete research anthology and summary could greatly benefit this field of research.

[20] See U.S. Government Accountability Office, "Bid Protest Annual Reports," webpage, undated(a), for a list of reports extending back to 1995.

bid protests. Most relevant to this study is a 2009 GAO report examining bid protest trends for DoD.[21] This study found that DoD bid protests at GAO were not at an all-time high, that the majority of protests were closed within 30 days, and that GAO did not need additional authority to dismiss protests characterized by some as "frivolous."

The Congressional Research Service has also published reports on bid protest activities.[22] It found that the number of protests (whether viewed as direct counts or relative to government contract spending) increased between FY 2001 and FY 2014 and that the percentage of sustained protests declined during this period. It found similar trends when it explored DoD protests, although the growth rate in protests was lower for DoD than for civilian agencies. The Congressional Research Service also examined companies' motivations in filing protests and the behaviors of agencies in light of the threat of a bid protest.

DoD has conducted internal studies on bid protests for defense procurements. The most recent report found that protest rates were increasing but sustained rates were stable.[23] The authors interpreted these trends as suggesting an increased propensity by bidders to protest rather than a decline in the quality of DoD procurements. Two reports on the DoD acquisition system also contain an analysis of bid protests.[24] In these reports, the authors note that the sustained rate for DoD protests at GAO have been relatively stable since 2008 and low (approximately 2 percent of protests), and they present an extensive analysis of protests by firm.

In 2013, Daniel I. Gordon explored the costs and benefits of the bid protest system and addressed some misperceptions. He concluded that the costs of the protest system were overstated and that the frequency of protests (relative to contract activity as a whole) was low. He also argued that the benefits of the protest system in terms of "transparency, accountability, education, and protection of the integrity of the U.S. federal acquisition system" outweighed its costs.[25]

Khoury, Walsh, and Ward (2017) found, among other things, that (1) the majority of protests were resolved within 30 days, (2) there were significant differences in the sustained rate between government agencies, (3) protesters that filed supplemental protests had a greater chance of seeing their protest sustained, and (4) how protest statistics are measured (e.g., relative to total actions, protester, or procurement) makes a difference.[26]

---

[21] U.S. Government Accountability Office, *Report to Congress on Bid Protests Involving Defense Procurements*, Washington, D.C., B-401197, April 9, 2009b.

[22] See Moshe Schwartz and Kate M. Manuel, *GAO Bid Protests: Trends and Analysis*, Washington, D.C.: Congressional Research Service, R40227, July 21, 2015, and Kate M. Manuel and Moshe Schwartz, *GAO Bid Protests: An Overview of Time Frames and Procedures*, Washington, D.C.: Congressional Research Service, R40228, January 19, 2016.

[23] Douglas J. Buettner and Philip S. Anton, *Bid Protests on DoD Source Selections*, Office of the Under Secretary of Defense for Acquisition, Technology, and Logistics, Performance Assessments and Root-Cause Analyses, June 2017.

[24] Office of the Under Secretary of Defense for Acquisition, Technology, and Logistics, *Performance of the Defense Acquisition System: 2015 Annual Report*, Washington, D.C., September 16, 2015; Office of the Under Secretary of Defense for Acquisition, Technology, and Logistics, *Performance of the Defense Acquisition System: 2016 Annual Report*, Washington, D.C., October 24, 2016.

[25] Gordon, 2013, p. 45.

[26] Khoury, Paul F., Brian Walsh, and Gary S. Ward, "A Data-Driven Look at the GAO Protest System," *Pratt's Government Contracting Law Report*, Vol. 3, No. 3, March 2017. The difference in sustained rates for supplemental protests is further explained as being more of an associative rather than causal trend. For example, a protester may expand a protest upon obtaining the agency record, and supplemental protests cannot occur if GAO dismisses the protest.

Hawkins, Yoder, and Gravier (2016), through a combination of modeling and surveys, explored how the fear of a protest influenced government procurement behavior.[27] The authors concluded that the fear of a protest "increases compromised technical evaluations, added procurement lead time, and transaction costs, while it decreases contracting officer authority and is associated with source selection method inappropriateness."[28]

Saunders and Butler (2010) explored bid protest outcomes at both GAO and COFC as a component of their research.[29] This study was unusual because there are generally more published reports on GAO protests than COFC protests. The authors found that about half of the protests filed with COFC had a prior GAO history and that the "vast majority of protests brought both to the GAO and COFC resulted in identical results." However, they noted a possible increasing trend in the protester losing at GAO and subsequently winning at COFC.[30]

Finally, RAND explored bid protests for the U.S. Air Force in two published reports.[31] The authors observed that the rates of protests relative to contract awards were small and declined between FY 1995 and FY 2008. They found that "the likelihood of bidders pursuing protests with GAO has been declining over time at a rate of between 8 and 9 percent per year, after controlling for other factors."[32] They also observed differences between Air Force contracting centers but found no difference between "protest outcomes and the commodity or service being acquired by the Air Force."[33]

Conclusions

In summary, the bid protest system for federal government acquisitions and procurements has evolved over the years. There are many, many scholarly resources on this topic. For the sake of brevity, this chapter provided a broad overview of the evolution of the federal government's bid protest system and general historical trends. More specifically, it explored the bid protest history and goals for DoD acquisitions and procurements, along with theories and quantitative analyses that explain the current process.

---

[27] Timothy G. Hawkins, Cory Yoder, and Michael J. Gravier, "Federal Bid Protests: Is the Tail Wagging the Dog?" *Journal of Public Procurement*, Vol. 16, No. 2, Summer 2016.

[28] Hawkins, Yoder, and Gravier, 2016, p. 152.

[29] Raymond M. Saunders and Patrick Butler, "A Timely Reform: Impose Timeliness Rules for Filing Bid Protests at the Court of Federal Claims," *Public Contract Law Journal*, Vol. 39, No. 3, Spring 2010.

[30] We note that the numbers are very small (a few protests each year), and, thus, it is difficult to say whether this trend is meaningful.

[31] Frank Camm, Mary E. Chenoweth, John C. Graser, Thomas Light, Mark A. Lorell, and Susan K. Woodward, *Government Accountability Office Bid Protests in Air Force Source Selections: Evidence and Options—Executive Summary*, Santa Monica, Calif.: RAND Corporation, MG-1077-AF, 2012; Thomas Light, Frank Camm, Mary E. Chenoweth, Peter Anthony Lewis, and Rena Rudavsky, *Analysis of Government Accountability Office Bid Protests in Air Force Source Selections over the Past Two Decades*, Santa Monica, Calif.: RAND Corporation, TR-883-AF, 2012.

[32] Light et al., 2012, pp. xii–xiii.

[33] Light et al., 2012, p. xiii.

CHAPTER THREE

Stakeholder Perspectives on the Bid Protest System

This chapter reviews the current U.S. bid protest environment and processes as perceived by U.S. government departments and agencies, by companies that have been or could be bidders, and by other stakeholders. It relies on discussions that we conducted and data that we obtained from relevant parties inside and outside government. Given the definition of bid protests in Chapter Two and the discussion of their history and legal record, it is not surprising that these parties have varying interpretations of the efficiency and efficacy of the system and of their respective roles and responsibilities.

FY 2017 NDAA Guidelines for Obtaining Stakeholder Perspectives

Section 885 of the FY 2017 NDAA specifically directed that this study include the collection and analysis of perceptions of the bid protest system from DoD employees and companies that have bid on contracts. More specifically, the language states,

> [Describe the impact of the bid protest system across DoD] for employees of the Department, including the contracting officers, program executive officers, and program managers, the extent and manner in which the bid protest system affects or is perceived to affect—
>
> (A) The development of a procurement to avoid protests rather than improve acquisition;
> (B) The quality or quantity of pre-proposal discussions, discussions of proposals, or post-award debriefings;
> (C) The decision to use lowest price technically acceptable procurement methods;
> (D) The decision to make multiple awards or encourage teaming;
> (E) The ability to meet an operational or mission need or address important requirements;
> (F) The decision to use sole source award methods; and
> (G) The decision to exercise options on existing contracts.
>
> [Describe the impact of the bid protest system on corporate decisionmaking,] with respect to a company bidding on contracts or task or delivery orders, the extent and manner in which the bid protest system affects or is perceived to affect—
>
> (A) The decision to offer a bid or proposal on single award or multiple award contracts when the company is the incumbent contractor;
> (B) The decision to offer a bid or proposal on single award or multiple award contracts when the company is not the incumbent contractor;

(C) The ability to engage in pre-proposal discussions, discussions of proposals, or post-award debriefings;

(D) The decision to participate in a team or joint venture; and

(E) The decision to file a protest with the agency concerned, the Government Accountability Office, or the Court of Federal Claims.[1]

To address this NDAA mandate, we held discussions with relevant employees from the Department of the Army, the Department of the Air Force, the Department of the Navy (including the Marine Corps), and the Defense Logistics Agency (DLA) to get their perspectives of the impact of the bid protest system on their procurement planning and execution processes for awarding contracts. In addition, we held discussions with representatives from various trade associations that represent a broad set of companies, both large and small, to get their perspectives of the impact that the bid protest system has had on corporate decisionmaking for bidding on potential DoD contracts. Finally, we held discussions with staff from private law firms who represent companies in the bid protest process to get their perspectives on the bid protest system. Appendix C reproduces the questionnaire that guided our discussions with relevant DoD employees.

DoD Stakeholder Perspectives

Our discussions with relevant DoD employees allowed us to develop a better understanding of the impact of the bid protest system on DoD procurements.

Overall, we found that while each military service tracks bid protests in a different manner, each uses GAO data as its primary source for tracking bid protests. Because the number of bid protests at COFC is much smaller, the services and DLA do not spend as much time tracking protests filed in that forum.

That said, we found a perception among the services that several bid protests submitted to COFC had previously been submitted to GAO, which either had found them to be unsubstantiated or had dropped them when the service decided to initiate corrective action.

The service personnel indicated that they did not track the relationship between changes in procurement funding and the number of bid protests. However, they reported anecdotally that there is a general belief that reductions in procurement dollars have affected the number of protests. In Chapters Four and Five, we examine protest rates relative to procurement spending and contracts awarded.

The services also reported that they did not track or collect data on whether companies are more or less likely to file a bid protest as a normal course of their business strategy. This was the case for DLA as well. Service personnel also indicated anecdotally that each company has a "unique course of business" and that, inasmuch as protests are situation-specific, it is difficult to determine a company's likelihood of filing a bid protest.[2] However, all the services agreed that there is a commonly held belief that a contractor is more likely to file a bid protest if it is an incumbent that has lost in a follow-on competition. The services gave a variety of reasons

---

[1]   Pub. L. 114-328, Section 885.

[2]   One service representative suggested that large companies are more likely to submit a bid protest after losing a competition because they have the resources to do so.

for an incumbent filing a bid protest—ranging from structuring an orderly transition of its workforce to obtaining a follow-on bridge contract from the government that would provide additional revenue.

We also asked service personnel whether they thought the specter of a bid protest influenced acquisition decisions in terms of how requests for proposals (RFPs) are structured and evaluated. They responded that acquisition decisions—specifically, the structure and evaluation mechanisms for RFPs—are influenced primarily by statutes (such as CICA, the Federal Acquisition Streamlining Act, the Small Business Act, and the Buy American Act), as well as various NDAA provisions that may or may not be codified in Title 10 of the U.S. Code. From the services' perspective, a bid protest is a process to ensure that the underlying statutes and regulations have been followed.

DoD and service contracting officers do not consider the prospect of receiving a bid protest to be a top priority as they develop RFPs. Their primary focus is on ensuring that requirements and evaluation criteria are clearly defined to minimize bid protests. When asked whether the fear of a bid protest would limit acquisition and contracting options, these personnel responded that they were not afraid of receiving a bid protest and that, in general, they believed that acquisition and contracting options were not being thwarted by the fear of bid protests.

However, contracting officers also noted that the possibility of a bid protest did affect the type of contract or contract vehicle they selected—usually prompting them to favor a price-related choice or existing task/delivery order–type contract if appropriate. They added that possible bid protests also affect the scrutiny that source-selection documentation receives from legal counsel, as well as the amount of time required to award the contract, which can result in programs missing key milestones or losing funds.

Service personnel noted that bid protests had a small impact on whether to use such approaches as LPTA or sole-source contracting. While each service had experience with LPTA contracting, most felt that it was appropriate only for less complicated or less technical procurements for which it is easy to determine whether the contractor is technically acceptable. The consensus was that LPTA can limit flexibility in awarding best-value contracts and is not feasible for complex or technically demanding procurements. With respect to sole-source contracting, the services felt that bid protests had no impact whatsoever. Personnel stated that CICA and exceptions to full and open competition are well understood by contracting officers.

They also stated that the potential for a bid protest did not affect their ability to meet operational and warfighting requirements. The potential for a bid protest also had little bearing on contracting officers' decisions to make multiple awards, encourage teaming, or exercise contract options. However, some contracting officers indicated that they were concerned that a bid protest would delay their ability to meet program contracting milestones and risk program funding reductions if they could not meet obligation and expenditure benchmarks.

We heard in our discussions that the services have made changes to improve the procurement and contracting process that may reduce the number of bid protests in the future. The Army has several ongoing initiatives, ranging from improving organization debriefings and training on source-selection procedures to improving industry exchanges and capturing lessons learned. These efforts include the following:

- Army Contracting Command–Aberdeen Proving Ground has established a red-team debriefing process through the Source Selection Center of Excellence to assist contracting teams with debriefs and improve the transparency of the procurement process.
- The U.S. Army Health Contracting Activity has started including a redacted copy of the award decision and other documents in the debriefing process to better explain the award to unsuccessful offerors in their multi-hundred-million-dollar indefinite delivery/indefinite quantity health care acquisitions.
- The 419th Contracting Support Brigade now requires all acquisition plans for actions exceeding $10 million to be reviewed by the principal assistant responsible for contracting, who also organizes robust solicitation and contract review boards with key stakeholders to ensure that the acquisition process is properly followed and documented.

Personnel from each of the services and DLA stated that exchanges with industry—such as holding industry days, conducting industry outreach (which often involves teaming with small business representatives), and soliciting feedback on draft RFPs—are methods they use to improve dialogue with companies and increase the transparency of the procurement process.

The Air Force relies on another initiative to dissuade unsuccessful offerors from filing bid protests: It gives unsuccessful offerors the opportunity to participate in extended or enhanced post-award debriefings.[3] In these debriefings, the Air Force provides an unsuccessful offeror's outside counsel with otherwise protected information to fully explain its decision, either to eliminate an offeror from a competitive range or to award the contract to another offeror. Most of the time, the information includes source-selection documents that address the unsuccessful offeror's complaints. These are documents that an unsuccessful offeror's outside counsel would receive under a GAO protective order if the offeror submitted a bid protest with GAO. With the Air Force offering its source-selection documents (or even its entire agency record) at this juncture, an offeror's outside counsel can ascertain in advance of filing a bid protest whether such a move is warranted and provide an opinion on the fairness, impartiality, and rationality of the award decision. The Federal Aviation Administration has adopted a similar process and has seen results similar to those of the Air Force, with extended debriefings frequently resulting in an unsuccessful offeror's counsel dissuading the company from filing a bid protest or advising it to withdraw a previously filed bid protest.

In summary, DoD personnel described a general dissatisfaction with the current bid protest system. The prevailing thought is that contractors have an unfair advantage in the contracting process by impeding timely awards with bid protests. They stated that the federal government allows too many "weak" allegations in a protest, that the contractor has too much time to protest, and that GAO takes too long to respond (i.e., from summary dismissal through the CICA stay process).

Despite their general dissatisfaction, DoD personnel also offered ideas to improve the bid protest system. Some of these are listed below. Note that these ideas are not official positions or proposals from DoD leadership.

- Reconsider the CICA stay process, with attention given to the override process.
- Add an acquisition value threshold for a contractor to protest above the contracting agency, such as task-order protest limitations.

---

[3]   Interagency Alternative Dispute Resolution Working Group, undated(a).

- Require the Office of Management and Budget (or another, more appropriate organization) to routinely publish GAO and COFC "lessons learned," such as when requests for summary dismissal might be successful.
- Create a statutory protective order scheme for agency-level protests whereby protesters can examine source-selection and proposal information to help resolve the protest.

Industry Stakeholder Perspectives

In parallel with discussions and data requests of DoD personnel, we solicited opinions and perceptions from trade associations and private law firms of the impact that bid protests have on their corporate decisionmaking. Overall, the private sector views bid protests as a healthy component of the acquisition process because protests hold the government accountable and provide information on how the contract award or source selection was made. It sees the bid protest system as providing transparency throughout the procurement process. Industry representatives also stated that if bid protests were not allowed or were curtailed, companies would likely make fewer bids.

Company and Trade Association Perspectives

The trade associations we contacted reported that they did not have a database for tracking bid protests. Instead, they indicated that their membership largely relies on internal and external legal counsel to remain up to date on developments in GAO and COFC bid protest case law.

Incumbency does not seem to matter in a company's calculations about whether to bid on a solicitation. Obviously, incumbent contractors have greater reason to believe that they provide the best service to the government. But respondents reported that their evaluations about whether to make a bid primarily hinged on estimates of their probability of winning and the cost of preparing proposals rather than on incumbency. Bid protests did not factor into their decisionmaking at this stage. Occasionally, companies said, they protested to learn more about why they did not win the contract.

The private sector's ability to engage with government representatives in pre-proposal discussions or post-award debriefings varied. Companies reported that they almost always attempted to engage in pre-proposal discussions. Those discussions influenced the companies' estimates of their probability of winning and, therefore, their decision about whether to bid. Some would not bid if they could not engage in proposal discussions. If there was an opportunity for post-award discussions, companies reported that they always participated because they viewed them as a good learning experience.

A company's decision about whether to participate in team or joint ventures was driven by the scope of the work in the solicitation. When a bid announcement specified a work scope that was broader than a contractor's existing capabilities, the contractor would seek teammates to increase the probability of winning the contract. The decision about which team member would serve as the prime contractor was subject to negotiation among the prospective partners. Ultimately, the prime selected was the candidate most likely to win the contract.

Companies reported that they decided to file a bid protest with GAO and COFC if they believed that there had been serious wrongdoing by the evaluators, if they lacked an understanding of why they lost, or if a simple cost-benefit analysis showed that filing a bid protest made sense. On the other side of the equation, companies also weighed the potential for "ill will" that could be created when they considered filing a bid protest. This was an important factor for large companies with a substantial number of contracts with the federal government. Large companies generally filed a bid protest only if they thought the government did not follow its source-selection procedures or that an error was made that was substantial enough to change the outcome.

Another concern in our discussions was the quality of post-award debriefings. The worst debriefings were characterized as skimpy, adversarial, or evasive and failed to provide reasonable responses to relevant questions. Debriefings that comply with FAR 15.505 and 15.506 often do not provide unsuccessful offerors with enough information to ascertain whether their proposals were evaluated properly. This also is the case for standard FAR-compliant debriefings, which usually provide limited feedback on the strengths, weaknesses, and deficiencies of a proposal. Moreover, standard FAR debriefings generally do not provide cohesive explanations for the government's evaluation conclusions and contract award decisions. This frustrates unsuccessful offerors and can lead them to speculate about the reasons they were eliminated from a competition or not awarded a contract. In desperation, unsuccessful offerors will submit a bid protest to obtain government documents that delineate the rationale for the contract award. The bottom line is that too little information or debriefings that are evasive or adversarial will lead to a bid protest in most cases.

The competency of the DoD acquisition workforce was another area with which private-sector representatives had concerns.[4] The consensus among the industry representatives we contacted was that the acquisition workforce is insufficiently staffed and could benefit from additional training. The workforce was cut massively in the 1990s and is still in the process of rebuilding. New process requirements are constantly being added or changed to meet the rapidly evolving marketplace. Future budgets are likely to severely constrain training, recruiting, and retention. The solution the industry representatives proposed was for DoD to structure, educate, and fund a workforce that is sufficient to meet the process and outcome requirements that are levied on it.

Organizational conflict of interest was yet another concern raised during our discussions. The issue, thought to pertain primarily to small businesses and contractors that provide services, concerns how the government determines when a conflict of interest exists and what solution is required to mitigate the conflict.[5] Because each potential conflict of interest arises from a unique set of circumstances, it requires significant judgment by the contracting officer and the government's legal counsel to resolve. Contractors reported that they were sometimes

---

[4] National Defense Industrial Association, *Pathway to Transformation: NDIA Acquisition Reform Recommendations,* Arlington, Va., November 14, 2014.

[5] Organizational conflicts of interest occur when activities or relationships with other entities mean that the institution is unable to render impartial assistance or advice to the government, cannot perform the federal contract work in an objective way, or has an unfair competitive advantage. Such conflicts could result when the nature of the work being performed on a federal contract creates an actual or potential conflict of interest for a future award, which could result in restrictions on that award. There are three basic categories of organizational conflicts of interest: *biased ground rules* (FAR 9.505-2; for example, preparing or writing specifications or work statements that are used in a funding opportunity); *impaired objectivity* (FAR 9.505-3; for example, evaluating or assessing the performance of products or services of others within the same organization); and *unequal access to information* (FAR 9.505-4; for example, gaining access to nonpublic information [e.g., budgets or budget information, statements of work, evaluation criteria] over the course of a federal contract).

frustrated by the overly restrictive requirements to resolve a conflict of interest, which could limit their opportunities for future work.

Outside Legal Counsel Perspectives

We held discussions with attorneys from several private law firms that had bid protests as part of their portfolios. These attorneys provide a unique perspective because they had represented both clients whose contract awards had been protested and clients who had submitted bid protests. They had also been involved in protests of large and small procurements, represented clients before GAO and COFC, and possessed a good understanding of the issues concerning the bid protest system.

The attorneys we contacted concurred that post-award debriefings can often be of poor quality and that these situations usually lead to bid protests. They suggested that more information needed to be provided during debriefings. They also contended that the administrative record should include everything. For example, GAO does not require that the contracting agency produce the full administrative record related to the bid and, indeed, heavily redacts the administrative record, which leads to suspicions about conclusions. The attorneys argued that GAO should require the agency to produce the complete administrative record. This is not an issue at COFC, where proceedings require the complete administrative record.

The attorneys we contacted also opined that reforms to the bid protest system should be guided by data, not emotion or anecdote. They noted that some empirical data exist today and that those data suggest caution before making drastic changes to the bid protest system or processes. They pointed to GAO's recent reports to Congress on the bid protest system.[6] For example, about half of all protests provide "effective relief," meaning that in almost half of all procurement protests, GAO either sustained the protest or the agency took corrective action to fix a flaw in the procurement. This rate suggests that the bid protest system provides a necessary oversight function that greatly improves the integrity and quality of procurements. It also suggests that the current acquisition system does not suffer from a flood of frivolous bid protests when compared with the vast amount of protests in which some sort of relief is granted.

The attorneys believed that bid protests were a healthy component of the acquisition process and that they guard against fraud and abuse. Because companies lack clear data on the root causes of bid protests and their impact, they concurred in calling for further in-depth studies on bid protests and potential causal factors so that changes to the bid protest system actually fix problems rather than merely address symptoms or other problems. The private sector understands congressional and DoD frustrations with bid protests and their impact on acquisition. Additionally, private industry representatives with whom we spoke opposed any legislative action in this area, such as Section 827 of the FY 2018 NDAA,[7] and believed that this provision would undercut the fundamental purpose of the bid protest system, which is to hold agencies accountable for following the law and solicitation procedures.

---

[6] For a collection of GAO's annual reports on bid protests, see GAO, undated(a).

[7] See Public Law 115-91, National Defense Authorization Act for Fiscal Year 2018, December 12, 2017. Section 827 is titled "Pilot Program on Payment of Costs for Denied Government Accountability Office Bid Protests."

Conclusions

Perspectives on the bid protest system varied greatly between DoD personnel and the private sector. DoD personnel expressed a general dissatisfaction with the current bid protest system. The prevailing thought was that contractors have an unfair advantage in the contracting process by potentially impeding timely awards with bid protests. They asserted that the federal government allowed too many "weak" allegations in a protest and that contractors had too much time to protest, delaying procurements. In contrast, private-sector representatives strongly supported the bid protest system because they viewed it as providing transparency to the contracting process and holding the government accountable for following the law and its own solicitation procedures.

CHAPTER FOUR

Quantitative Analysis of DoD Bid Protest Activity Since FY 2008 at GAO

Section 885 of the FY 2017 NDAA calls for an analysis of the history of DoD bid protests and associated trends. As we noted in Chapter One, it was not possible to address this request in its entirety because of the lack of available data. In this chapter, we present the quantitative analysis that was possible based primarily on bid protest histories from GAO. We present a similar analysis for COFC in Chapter Five. In both chapters, we used data compiled and provided by the respective venues that covered the period from roughly 2008 through the end of 2016.[1]

In this chapter and in Chapter Five, we begin by discussing some general characteristics of the data provided. We subsequently describe the major trends and characteristics of the data in a topical fashion rather than point by point as articulated in Section 885. We believe that organizing the content in this way makes each chapter easier to follow and allows us to discuss data trends in a way that may be more useful to decisionmakers who are thinking broadly of improvements to the bid protest system.

GAO Data Characterization and Issues

To support this congressionally mandated study, GAO provided a record of bid protest activity for protests with that organization starting in FY 2008 through the end of FY 2016. Derived from GAO's docket system, the record contained 21,186 actions related to protests, including actual protests, reconsideration requests, and requests for entitlements and costs. It covered all government agencies under GAO's protest jurisdiction. Each record included various characteristics of the protest: the primary contracting agency, protester name, dates the case was filed and closed, protest disposition, approximate value of the procurement (as reported by the protester), and whether the protester was a small business (again self-reported).[2]

An important point to mention is how GAO records protests in its docket system. When a new protest is filed against a procurement (either pre- or post-award), GAO creates what is known as a *primary B number*, a unique, numerical identification for that procurement. Protest actions are specific events associated with a primary B number. These actions are given what is referred to as a *dot number*, which is attached to the primary B number. Therefore, the initial protest filing has the format [primary B number].01 (e.g., B400153.01), where 01 is the dot

---

[1] The GAO data covered FY 2008 through the end of FY 2016. The COFC data covered the start of calendar year (CY) 2008 through mid-2017. We chose the 2008 starting point because it was after the change that added the review of task-order protests at GAO. This point avoids any discontinuity or needed normalizations in analyzing time trends prior to 2008.

[2] Note that the record was not comprehensive. Nevertheless, this chapter explores various characteristics from the data and how those data differentiate between protest activities or outcomes.

number. As a protest case evolves, subsequent dot numbers may be added (e.g., B400153.02) for such actions as amended or additional claims, a new party protesting, or a request for entitlement. Typically, most procurements that are protested have one or two associated dot numbers. However, large, complex cases might have tens of associated dot numbers (the highest in the data set was 53). For ease of discussion, we refer to an individual dot number record as a *protest action*.[3]

The fact that a procurement might have multiple records makes the interpretation of related trends and statistics complex.[4] This analysis of protest actions examines protest activity and workload at GAO. GAO determines the disposition of cases at the protest action level but not at the procurement level. Thus, any given procurement could have multiple outcomes, depending on the specifics of the individual protest actions.

GAO generally reports statistics at the protest action level. A disadvantage of this approach is that procurements with multiple actions might be overrepresented in averages (which is true for larger, complex procurements) and an analysis may overcount problematic procurements. As an alternative, some researchers analyze protest activity at the procurement level (primary B number). This approach is useful in understanding issues at the procurement level (e.g., how many procurements have experienced protests or required some type of corrective measures). But examining protest outcomes becomes problematic with this approach because outcomes across actions can be mixed. For our analysis, we generally followed GAO's approach of analyzing bid protest statistics—particularly for protest outcomes. For basic trends and simple statistics, we sometimes report values both ways or focus on the procurement level (particularly when we discuss protests relative to overall contracting levels).[5]

The disposition of a protest action can have several outcomes: sustained, denied, granted,[6] dismissed, or withdrawn. However, these outcomes do not fully indicate when the protester received some form of relief. An agency may take voluntary action ahead of GAO's decision. This action is called *corrective action* and results in either GAO dismissing the case or the protester withdrawing the protest. To better track when protesters are successful, GAO uses two metrics: sustained rate and effectiveness rate. The sustained rate is the number of actions for which GAO sustains the protester's claim divided by the number of protest actions that go to decision.[7] The effectiveness rate is the number of protest actions that are either sustained or are subject to corrective action relative to *all* protest actions. We will use these same metrics to measure protest outcomes. However, we calculated the sustained rate relative to the *entire*

---

[3] We included reconsiderations as part of protest actions but did not include entitlement requests. Because successful reconsiderations are rare, some observe that their inclusion biases the protest statistics lower (see Schwartz and Manuel, 2015). We report basic summary statistics for bid protests only in Appendix A. We note that reconsiderations are infrequent and that basic averages of protest outcomes are not significantly changed when including reconsiderations.

[4] For a good discussion of these issues, see Khoury, Walsh, and Ward, 2017, and Schwartz and Manuel, 2015.

[5] We used an inverse weighting method to analyze these data at the procurement level. For each primary B number, we calculated the number of associated actions (dot numbers). Each action is inversely weighted to that total. For example, if a procurement had four associated actions, then each of those actions would be weighted 0.25 in any averaging or analysis. This approach allowed us to avoid making judgments about the characteristics of the protest for mixed cases, such as when one of the protesters was a small business and the others were not.

[6] This disposition is applicable to reconsideration requests found in favor of the protester.

[7] These protest outcomes are either sustained or denied and are also referred to as *merit cases* because the case is not dismissed or withdrawn.

population of protest actions and not just those that went to GAO decision. This change kept the effectiveness and sustained rates comparable.

GAO Time Trends

One important issue to explore is the trend in protest activity over the past few years. In Figure 4.1, we display the number of protest actions from FY 2008 through FY 2016.

There has been a significant upward trend in protest activity at GAO between FY 2008 and FY 2016, a period in which activity for both DoD and non-DoD agencies has approximately doubled. Protest actions associated with DoD agencies accounted for roughly 60 percent of the total protest actions over this period. Even excluding task-order protests (which were added to GAO's jurisdiction in FY 2008), the upward trend is still statistically significant.[8]

To put this trend in context, one needs to go back further in time. To do this comparison, we supplemented GAO's data by adding prior years based on a 2009 GAO report on DoD protest activity.[9] Figure 4.2 shows the number of procurements protested (a sum of primary B numbers) between FY 1989 and FY 2016. The blue bars show data from the 2009 GAO report; the red bars are based on the data set that GAO provided to RAND. There is only one year of overlap (FY 2008) between the two data sets. The values do not match exactly but are within 4 percent of each other. Nonetheless, the time trend clearly shows that protest activity was much higher in the late 1980s and early 1990s than it is today.

Figure 4.1
Protest Actions at GAO, FYs 2008–2016

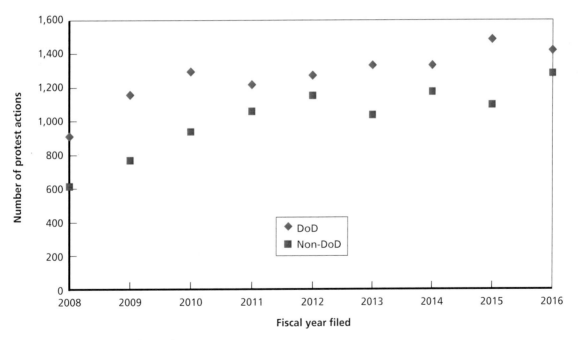

SOURCE: RAND analysis of GAO data.
RAND *RR2356-4.1*

---

8   See Appendix A, Figure A.5.

9   GAO, 2009a.

Figure 4.2
DoD Procurements Protested at GAO, FYs 1989–2016

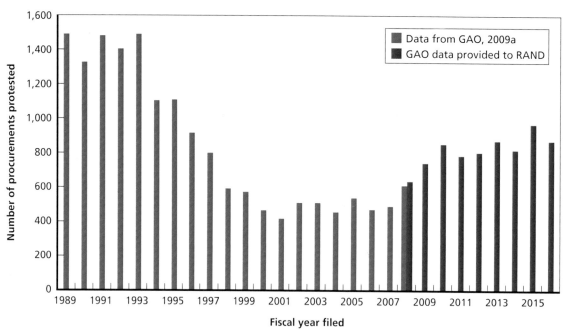

SOURCE: RAND analysis of GAO, 2009a, and GAO-provided data.
RAND RR2356-4.2

The previous time trends are indifferent to changes in DoD spending and contracting. To better understand whether spending and contracting changes are driving the trend, we obtained data on DoD contracts (numbers of contracts and contract dollars) from FPDS-NG.[10] Both the number of contracts and contract spending declined from FY 2008 to FY 2016.[11] This is counter to the trend for DoD bid protests. In Figure 4.3, we show the percentage of contracts protested and the number of procurements protested per billion dollars of spending. The increases of nearly 100 percent are statistically significant trends—measured either in terms of the percentage of contracts protested or in terms of the number of contracts protested per billion dollars. Still, it is important to note that the overall percentage of contracts protested is very small, less than 0.3 percent. This small value implies that bid protests are exceedingly uncommon for DoD procurements.

GAO Protests by DoD Agency

There are differences in the protest activity and trends between the DoD agencies. We segregated the DoD agencies into five groups: Department of the Army, Department of the Air Force, Department of the Navy (which includes the Marine Corps), DLA, and other DoD (all

---

[10] Appendix B provides more detail on this FPDS-NG analysis.

[11] Shown in Appendix A, Figure A.1.

Figure 4.3
Percentage of DoD Procurements Protested and Protests per Billion Dollars at GAO

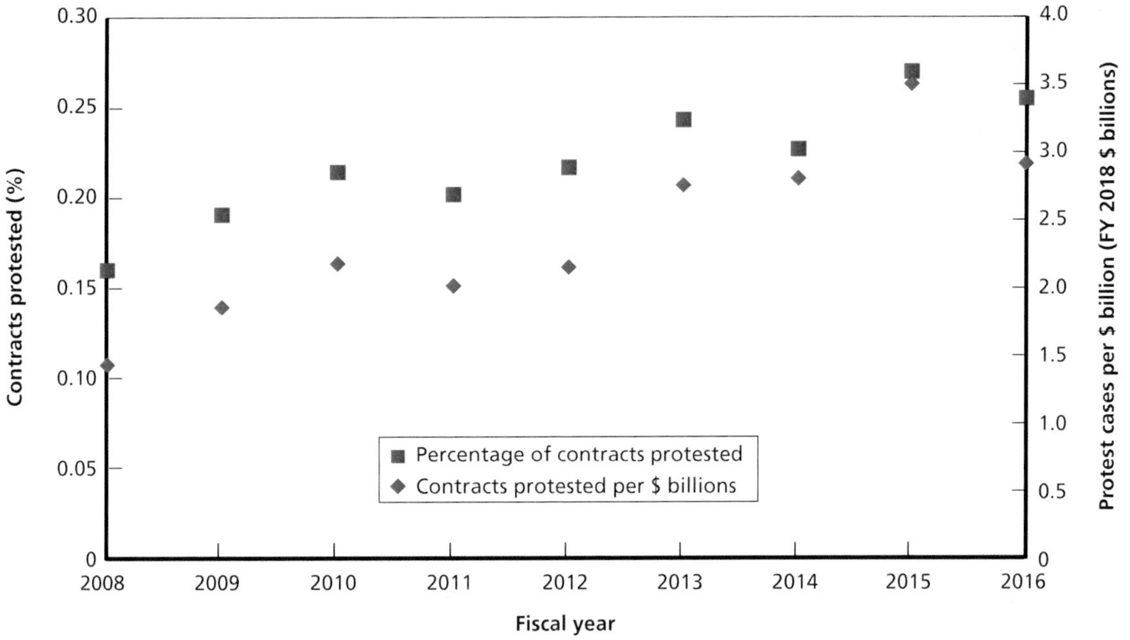

SOURCE: RAND analysis of GAO and FPDS-NG data.
RAND RR2356-4.3

remaining DoD agencies).[12] Figure 4.4 shows the number of protest actions in FYs 2008–2016 by agency. Notice that the Army has significantly more protest activity than the other services, but it peaked around FY 2010 and has declined since then. Activity for other DoD agencies, while generally lower, has increased since FY 2008.

To better contextualize differences, Table 4.1 compares the relative level of protests, spending, and contracts for each of the five DoD agency types. Presumably, a higher share of the contracting activity (dollars or numbers) should result in a greater share of protests, all other things being equal, if protests are a random fraction of all contracting activity (akin to an error rate). The percentage of contract spending correlates better with the percentage of protest actions than does the percentage of contracts (0.86 versus 0.20). The relative level of contract spending does order the agencies correctly with respect to protest activity (e.g., the Army is highest, the Navy is next highest), but it does not fully explain the differences. For example, the Army has relatively more protests compared with its spending or number of contracts. The higher proportion of protests might be due to the nature and type of goods and services (for example, the Army was the executive agent for Operation Enduring Freedom and Operation Iraqi Freedom). The Navy's protest levels were lower compared with its relative share of contract spending. The reasons for these differences between DoD agencies might be an area for future research.

---

[12] We originally started with just the three services and all other protests grouped into an "other DoD" category. But when examining the data more closely, DLA had a substantial number of protests (more than the remaining agencies combined). Thus, we split DLA into a unique grouping. As shown later, DLA's protest outcomes are very different as well.

Figure 4.4
GAO Bid Protest Actions, by Agency, FYs 2008–2016

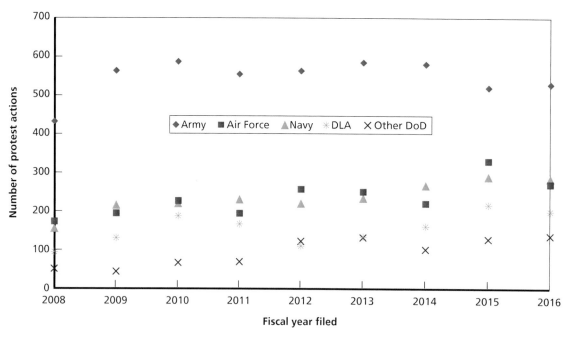

SOURCE: RAND analysis of GAO data.

RAND RR2356-4.4

Table 4.1
Percentage of Protests, Spending, and Contracts by DoD Agency at GAO, FYs 2008–2016

Agency	% of DoD Protest Actions	% of DoD Protested Contracts	% DoD Contract $	% DoD Contracts
Army	43	41	34	25
Air Force	18	18	19	9
Navy	19	19	28	19
DLA	12	16	11	44
Other DoD	8	7	9	3

SOURCE: RAND analysis of GAO and FPDS-NG data.

NOTE: Columns may not sum to 100% due to rounding.

Firms with Largest Contracts Awarded

An interesting comparison to make is whether the increasing protest trend also holds for the firms with the largest amount of funds awarded by DoD. In Figure 4.5, we show the aggregate number of protest actions from FY 2008 through FY 2016 for the top 11 firms (by FY 2016 revenue).[13] The firms were (in descending order of funds awarded) Lockheed Martin Corp.,

---

[13] The ranking came from Federal Procurement Data System: Next Generation, "Top 100 Contractors Report, Fiscal Year 2016," spreadsheet, undated.

Figure 4.5
Protest Actions by Top 11 Firms at GAO, FYs 2008–2016

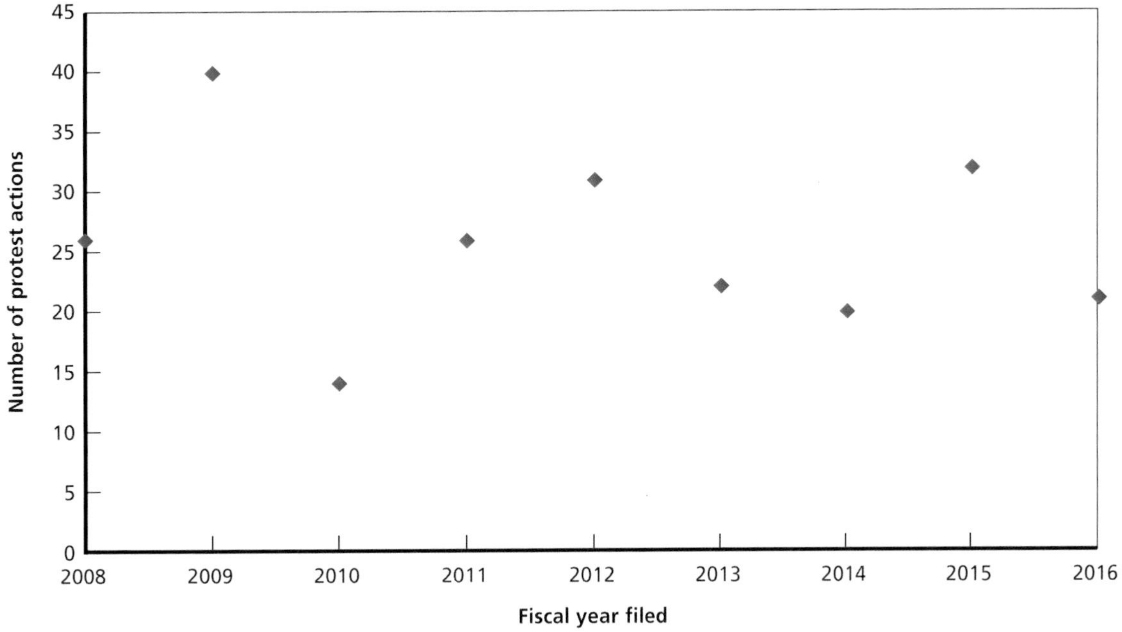

SOURCE: RAND analysis of GAO data.
RAND RR2356-4.5

the Boeing Company, Raytheon Company, General Dynamics Corp., Northrop Grumman Corp., United Technologies Corp., BAE Systems Plc., L3 Technologies, Huntington Ingalls Industries Inc., Humana Inc., and Bechtel Group Inc., comprising nearly 42 percent of total obligated dollars in FY 2016.

Figure 4.5 shows that protest activity by these 11 firms has remained relatively constant and may be slightly declining. (Due to the variability, the time trend is not statistically significant.) It has been speculated that one of the causes of the recent increase in the level of bid protest activity is the trend of reduced government (and DoD) spending over this same period. As spending declines, firms are thought to be more likely to protest to potentially win business for a declining revenue base in a more competitive environment.[14] These largest firms do not follow this trend in that their level of protest activity is not increasing as budgets decline. However, we have not fully explored how the funding trend for these firms has tracked over time, and a more thorough examination might yield better insights to this trend.

Pattern of Protest Filings at GAO

Bid protests are not filed at GAO uniformly throughout the year. Figure 4.6 shows the number of protest actions filed by month from FY 2008 to FY 2016. Notice that the filings peak around the end of the fiscal year and then drop sharply in November, December, and Janu-

---

[14] See, for example, Christian Davenport, "With Budget Tightening, Disputes over Federal Contracts Increase," *Washington Post*, April 7, 2014.

Figure 4.6
DoD Bid Protest Filings by Calendar Month at GAO, FYs 2008–2016

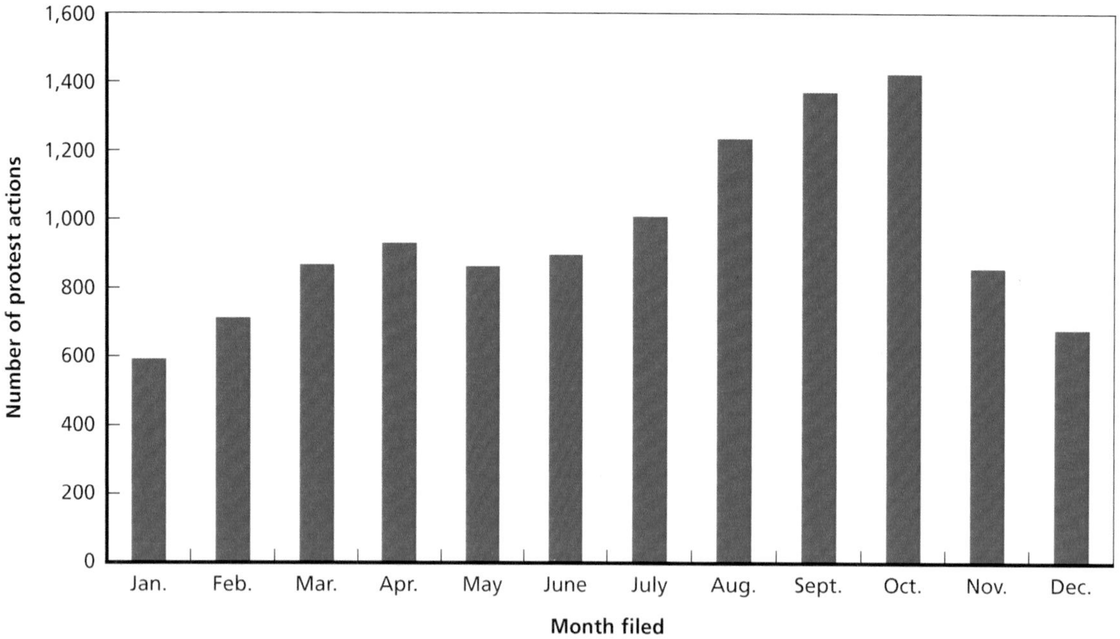

SOURCE: RAND analysis of GAO data.
NOTE: The numbers of filings are cumulative across FYs 2008–2016.
RAND *RR2356-4.6*

ary. This pattern has important implications for any effort to reduce GAO's decision timeline. Currently, GAO has 100 days to resolve cases. This 100-day window has the advantage of allowing GAO to smooth the workload between the peak and minimum filing periods (which are adjacent). There had been some discussion of potentially reducing GAO's timeline to 65 days.[15] This reduction, among other things, would provide less flexibility to GAO in managing its protest workload and may require additional staff to meet peak filing demand.[16] We suggest that the implications of this filing pattern be considered when exploring reductions to GAO's decision timeline.

Protest Characteristics

As noted earlier in this chapter, GAO tracks in its docket system a number of characteristics associated with a procurement, such as the approximate acquisition value, whether the protester is a small business, and whether the protest occurs pre- or post-award. In Table 4.2, we summarize these characteristics by both protest action and procurement.

A number of interesting features relevant to policymaking are evident from the averages in the table. Perhaps most striking is the fact that protests from small businesses account for

---

[15] See for example David Yang, "Senate Proposes Major Overhaul to the GAO Bid Protest Process," *Government Contracts Navigator*, October 3, 2017.

[16] GAO decision timelines are typically 90–100 days when a written decision is necessary.

Table 4.2
DoD Bid Protest Characteristics at GAO, FYs 2008–2016

Characteristic	All Actions	Weighted by Procurement
Observations	11,459	7,368
Average number of protesters	1	1.2
From small businesses[a]	53.1%	58.1%
Value under $0.1 million[a]	7.9%	10.5%
Task-order protests	10.6%	9.3%
Stay override issued	1.4%	1.3%
Alternative dispute resolution used	5.0%	3.6%
Protective order issued	48.9%	39.2%
Pre-award protest	26.9%	29.1%

SOURCE: RAND analysis based on GAO data.

[a] Values were self-reported by the protester.

more than half of all the protest actions.[17] The ratios of 53 percent or 58 percent are roughly consistent with the fraction of DoD contracts going to small businesses (about 65 percent) but are much higher than the fraction of DoD *contract dollars* going to small businesses (around 15 percent).[18] Furthermore, the fraction of bid protest actions by small businesses increased slightly from FY 2008 to FY 2016. The implication is that any changes or improvements to the bid protest system need to account for small businesses. Improvements aimed only at larger firms would miss the majority of DoD bid protest actions at GAO. As we observe in Chapter Five, small-business protests also form the majority of cases at COFC.

Another interesting feature of Table 4.2 is that 8–10 percent of protest actions are associated with procurements valued at under $0.1 million—with some falling under the definition of micro-purchases.[19] An interesting policy question is whether the cost to the government to adjudicate protests exceeds the value of the procurement itself (and is thus cost-effective).[20] Unfortunately, we were unable to find any government data associated with the costs to process bid protests.

---

[17] A caveat here is that protesters self-reported whether they were small businesses. Therefore, it is unclear whether all the firms identified as small business actually qualify under the formal definition of the Small Business Administration. Nonetheless, the majority of protesting firms self-identify as small businesses.

[18] See Appendix A, Figure A.3.

[19] Again, values are reported by the protester. Micro-purchases are those of $3,500 or less over the time frame of the protest data (see question 2 at Office of the Secretary of Defense, Defense Procurement and Acquisition Policy, "Government Purchase Card (GPC) Frequently Asked Questions (FAQs)," webpage, last updated October 17, 2017).

[20] Cost-effectiveness may not be the most important criterion in adjudicating a bid protest. Transparency of government spending might be the overriding consideration and, thus, cost-effectiveness is secondary. Again, these are large policy questions that are raised by the data.

Table 4.2 shows how infrequently a DoD agency issues a stay override—for less than 2 percent of procurements.[21] If bid protests substantially affect the procurement of urgently needed goods and services, one might expect this percentage to be higher. This low percentage raises a number of questions for policymakers. Is it too difficult for an agency to justify a stay override so that it is only used in very unusual circumstances? Are the agencies overly conservative when issuing overrides? Are most protests for procurements of an item or service that is not urgently needed? Unfortunately, the data do not provide further insight into these issues. Further research in this area might help determine why these rates are low.

DoD Bid Protest Outcomes at GAO

Table 4.3 summarizes protest outcome measures for DoD protests at GAO. The sustained rate is relatively small (as measured against all protest actions) but is over 12 percent when measured against actions that go to decision (merit cases). The majority of the relief to protesters takes the form of corrective action. The two outcomes combine into an effectiveness rate of approximately 40 percent. Stated another way, roughly 40 percent of all protest actions result in some change to the initial procurement decision or terms.

Perhaps even more striking is the stability of the effectiveness and sustained rates over time. Figure 4.7 shows trends from FY 2008 to FY 2016. The rates have been steady and may be slightly increasing with time. The upward trends for both rates are not statistically significant. However, if protests from one company that GAO debarred for a period of time due to excessive numbers of protests are excluded, the trends are significantly upward (although at a

Table 4.3
DoD Bid Protest Outcome Measures at GAO, FYs 2008–2016

Outcome Measures	All Actions	Weighted by Procurement
Observations	11,459	7,368
Sustained rate	2.6%	1.4%
Merit cases[a]	21.2%	NA[b]
Sustained rate for merit cases	12.2%	NA[b]
Corrective action rate	38.4%	38.6%
Effectiveness rate	41.0%	40.0%
Sustained rate (excluding reconsideration)	2.7%	1.5%
Effectiveness rate (excluding reconsideration)	42.4%	40.8%

SOURCE: RAND analysis of GAO data.

[a] A merit case is a bid protest action that goes to decision (GAO either sustains or denies the protest and a written decision is issued.) We report the sustained rate for merit cases to be consistent with GAO's approach.

[b] Weighting is inappropriate for a subset of the data.

---

[21] A stay override occurs when an agency overrides the automatic hold of execution of a contract (award or performance) during a protest at GAO. There are two justifications for such an override: (1) urgent and compelling circumstances or (2) performance of the contract is in the best interests of the United States. (See Schwartz and Manuel, 2015, and U.S. Army, *Override of CICA Stays: A Guidebook*, version 3, June 2008.)

Figure 4.7
Effectiveness and Sustained Rates for DoD Protests at GAO, FYs 2008–2016

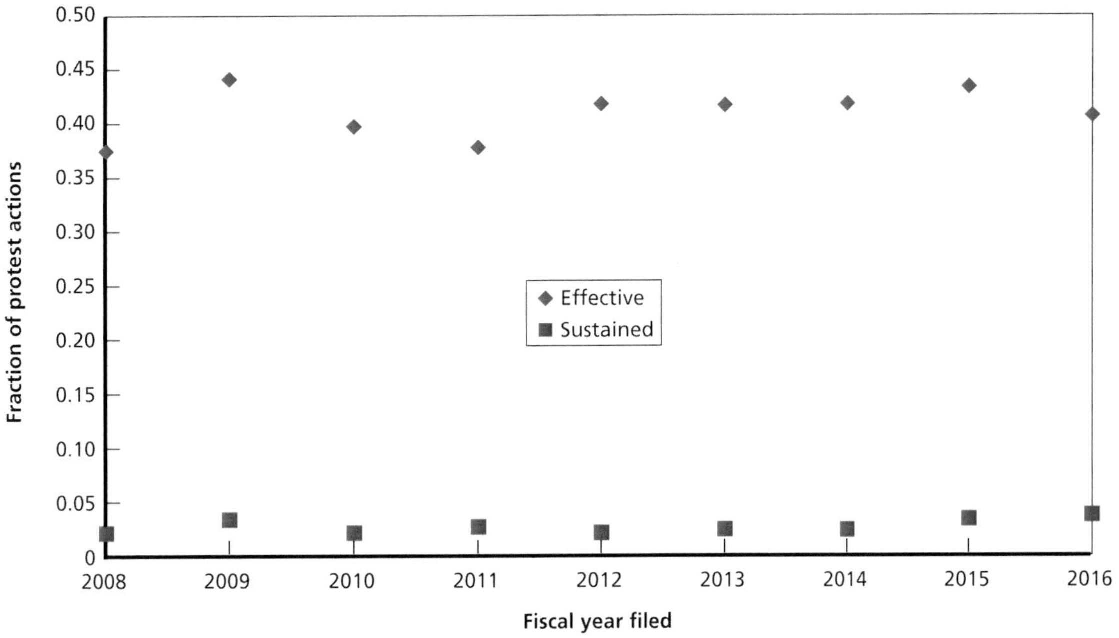

SOURCE: RAND analysis of GAO data.
RAND *RR2356-4.7*

very modest rate).[22] Nonetheless, the stability (or slight increase) in the effectiveness rate while the number of protests is increasing refutes the claim that meritless (some use the term *frivolous*) protests account for those increases.[23] If the increases were due to such protests, the effectiveness rate should be falling, which it is not. One possible explanation is that firms are more willing to protest procurements, where, before, they refrained from doing so due to either the costs involved or the potential for creating ill will with the customer.

One thematic line of inquiry in Section 885 of the FY 2017 NDAA involves an exploration of how protest outcomes are correlated with different characteristics of the procurement and protest. While there are several specific requirements for analysis, we explored a broad series of factors to see whether effectiveness or sustained rates changed with these features. To perform this analysis, we used logit regression to model both the effectiveness and sustained rates.[24] Our purpose was not to build predictive models but, rather, to identify differences between protest actions. In fact, the models we built have very poor predictive power. However, they identify significant differences in the broader population—some of them strong.

---

[22] In 2016, Latvian Connection was suspended by GAO from filing protests for one year for abusing the protest system by filing more than 150 protests in one year. See, for example, Steven Koprince, "150 Protests and Counting: GAO Suspends 'Frequent Protester,'" *SmallGovCon*, August 22, 2016.

[23] Several people in the private sector made this point to us during our discussions, and others have commented on the issue. See, for example, Alex Levine, "While Government Spending Is Down, Bid Protests Are Up," blog post, PilieroMazza, September 18, 2015.

[24] A discussion of logit regression is beyond the scope of this report. For more information, see StataCorp LLC, "logit— Logistic Regression, Reporting Coefficients," undated.

This lack of predictive power led us to an important broad observation about the bid protest system, which many stakeholders and participants stated to us during the course of this study: *The details of a protest case matter in terms of outcome.* It is not possible to predict the outcome of any case based on its general characteristics (e.g., the agency involved, the value of the procurement). Similarly, one cannot generalize about protest outcomes based on a single procurement protest.

Table 4.4 shows correlations between protest characteristics and effectiveness and sustained rates.[25] A "+" indicates a positive correlation (a higher rate), a "–" indicates a negative correlation (lower rate), and a "0" indicates no statistically meaningful correlation. Two repeated symbols indicate that the correlation was strong (a relatively large coefficient).

Table 4.4
Correlation of Protest Characteristics with Protest Outcomes at GAO

Characteristic	Effectiveness Rate	Sustained Rate
Task order	+	+
Small business	–	–
Top 11 firm (by awards in FY 2016)	+	+ +
Procurement value	0	+ +
Army procurement	+	– –
Air Force procurement	+	–
Navy procurement	0	–
DLA procurement	–	– –
Other DoD procurement	0	0
Pre-award protest	+	–
Solicitation type (e.g., RFP, RFQ)	+ (RFP, RFQ)	+ + (RFP)
Alternative dispute resolution	++	– –
Reconsideration	– –	– –
Number of protesters	+	+
Stay override issued	–	+ +
Protective order issued	+	+
Fiscal year	0	0
Initial filing	+	–

SOURCE: RAND analysis of GAO data.

NOTES: + + = strong, positive correlation, + = positive correlation, 0 = no correlation, – = negative correlation, – – = strong, negative correlation. All correlations that are not "0" are statistically significant. See Appendix A for details on the logit analysis and how we characterized "strong" correlations. Small-business status was self-reported by the protester. Other DoD agencies were set as the baseline value and show no correlation, by definition. RFQ = request for quote.

---

[25] Logistic regression details are shown in Appendix A.

A number of interesting correlations are evident in Table 4.4. The first is that task-order protests have slightly higher effectiveness and sustained rates compared with the rest of the protests. This positive, although small, correlation suggests that restricting protests on task orders is not supported by outcomes.[26] Small-business protesters tend to have lower sustained and effectiveness rates. Associated with this correlation is that small-business protests are 1.5 times as likely to have their protests dismissed for being "legally insufficient." However, if one looks at cases with protective orders,[27] there is a positive correlation with both outcomes (large and small businesses). These trends suggest that more protests filed by small businesses might be successful with better legal representation.[28] As we will see for COFC cases, where legal counsel is required, the effectiveness rate for small businesses is the same as for other businesses. The top 11 firms have higher effectiveness and sustained rates than the rest of the sample—suggesting that they are possibly more selective in the protests they file and spend more resources developing their cases. However, these rates have been declining with time (see Appendix A, Figure A.8), which may indicate that these firms have recently become less selective in their protests.

The DoD agencies have differing outcomes with respect to their procurements. Other DoD agencies were fixed at zero (set as the baseline value) and thus showed no correlation, by definition. The Army has a slightly higher effectiveness rate but a much lower sustained rate than the baseline. This difference in outcomes suggests that the Army is more aggressive in pursuing corrective action. However, some difference might be due to the nature of its procurements. The Air Force has a similar trend, although the sustained rate correlation is not as strongly negative. The Navy has a neutral effectiveness rate and a slightly lower sustained rate compared with the baseline. DLA has negative correlations for both and a particularly strong negative correlation with its sustained rate. It is unclear why these correlations exist for DLA; it may be due to that agency's processes or the nature of its procurements.

Reconsiderations rarely succeed. There was only one successful reconsideration case in our data sample. Pre-award protests have a slightly higher effectiveness rate but lower sustained rate, implying that most relief takes the form of corrective action. In terms of procurement type, protests on RFPs are correlated with higher sustained and effectiveness rates; RFQs are correlated only with higher effectiveness rates. Alternative Dispute Resolution (ADR) in bid protests appears to achieve its aim of expeditiously resolving protests by encouraging corrective action where appropriate and thereby results in fewer decisions to sustain. The number of protesters is positively correlated with increases in both rates. However, the number of protesters is highly correlated with the contract value, so the trend is not conclusive.[29] A stay override also behaves as one would expect. If an agency issues a stay override, it is unlikely to take corrective action because of a compelling reason to move forward with the procurement. More of the protest resolution falls to a decision (hence the higher rate). Yet, overall the net effectiveness rate is

---

[26] By *task-order protest*, we mean a protest of a specific task or delivery order award and not the underlying contract (e.g., indefinite delivery/indefinite quantity).

[27] Per GAO's bid protest guide (GAO 2009a), "If the record in a protest contains 'protected' information, that is, a company's proprietary or confidential data or the agency's source-selection-sensitive information, that information cannot be made public. In order to allow limited access to protected information relevant to a protest, GAO may issue a protective order."

[28] Protesters must use legal counsel under protective orders.

[29] See Appendix A, Figure A.6.

still much lower than typical, suggesting that DoD agencies are cautious when invoking overrides. There are no meaningful trends with regard to fiscal year (except as described earlier).

A specific analysis that Section 885 required was an examination of outcomes of task-order protests as a function of value. Specific value ranges were defined in Section 885. Unfortunately, the GAO docket system only records the procurement value in a range and not as a specific value. Thus, we kept to GAO's ranges in this exploration. Figure 4.8 shows the effectiveness rate of task-order protests as a function of value. As the complete data (represented by the blue bars) show, there is a marked increase in the effectiveness rate for bid protests of around $10 million. Not surprisingly, this is the threshold at which GAO had jurisdiction over DoD task-order protests during the FY 2008–FY 2016 time frame. Under specific circumstances, GAO will review task-order protests below this threshold. But these circumstances involve issues like the proposed scope of work being outside the initial contract.[30] Protests that are filed with GAO that are outside its jurisdiction are recorded in the docket system. If we remove task-order protests over which GAO had no jurisdiction, then there is no apparent difference in effectiveness rate with value (as shown by the red bars).

Similar to the trend for task orders, the broader protest population shows no correlation between the effectiveness rate and procurement value. However, there is a strong correlation with the sustained rate and value. Figure 4.9 shows the increase in the fraction of cases sustained by value. Multiple reasons may account for this trend: DoD agencies may be less

Figure 4.8
Effectiveness Rate at GAO by Task-Order Value, FYs 2008–2016

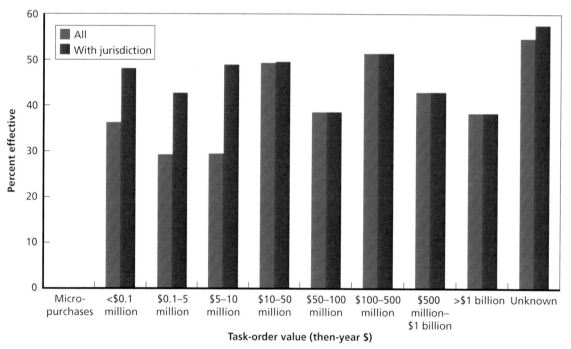

SOURCE: RAND analysis of GAO data.

RAND RR2356-4.8

---

[30] See, for example, Jay Carey and Kevin Barnett, "GAO's Task Order Protest Jurisdiction Expires Today," *Inside Government Contracts*, September 30, 2016.

Figure 4.9
Percentage of Cases Sustained at GAO by Procurement Value, FYs 2008–2016

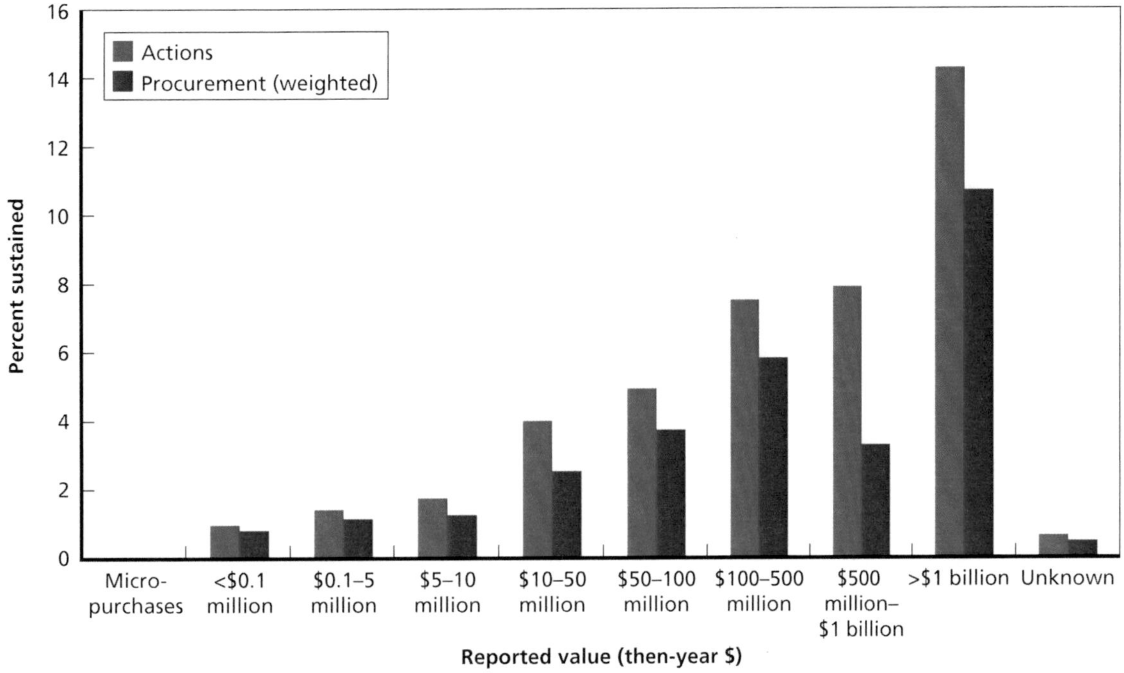

SOURCE: RAND analysis of GAO data.
RAND RR2356-4.9

likely to take corrective action on larger procurements, larger/more complex procurements may attract more protest actions and protesters (and, thus, errors are more likely to surface), and protesters may put more effort into protests involving larger procurements. Nonetheless, with the general sustained rate being very low, the number of actions/procurements is small. Therefore, some caution must be exercised in interpreting trends in the sustained rate.

Another important difference shown in Figure 4.9 is the low sustained rate for procurements with an unknown value. This is likely related to the earlier observation that pre-award protests have a lower sustained rate. Approximately, two-thirds of the pre-award protests have an unknown value, which is to be expected because there has not been a contract award.

Another analysis specifically called out in Section 885 of the FY 2017 NDAA was the following:

> An analysis of those contracts with respect to which a company files a protest (referred to as the "initial protest") and later files another protest (referred to as the "subsequent protest"), analyzed by the forum of the initial protest and the subsequent protest, including any difference in the outcome, between the forums.[31]

To address this issue (in the context of GAO protests), we examined the sustained and effectiveness rates between the *initial filing (action)* and all subsequent protest actions. Initial protests have a higher effectiveness rate but a lower sustained rate (this correlation was shown in Table 4.4). This difference is consistent with the view we discuss later in this chapter that

---

[31] Pub. L. 114-328, Section 885, para. 12.

corrective action is more likely initially and less likely with subsequent protests during the overall protest process. Subsequent protests typically arise *after* the agency report is filed. Therefore, the protester has an opportunity to refine its argument with additional protest actions and more information. Thus, subsequent protest actions have a higher sustained rate. This correlation does not correct for cases in which there are multiple protesters. If we restrict the sample to protests with only one protester, the same trends hold: Subsequent protests have a higher sustained rate.

GAO Bid Protest Timelines

An additional area of interest in Section 885 of the FY 2017 NDAA is the timeline for bid protest decisions. In Figure 4.10, we reproduce GAO's bid protest timeline from its website. Once a protest is filed at GAO, the corresponding agency has 30 days to file a report "responding to the protest, including all relevant documents, or portions of documents, and an explanation of the agency's position."[32] The protester then has ten days to respond to the agency's report. If the protest makes it this far, GAO may issue a decision within 100 days of the date the protest was initially filed. However, it is important to note that GAO may issue a decision to sustain or deny a protest or to dismiss it at any time during the process. Similarly, an agency may take corrective action before GAO resolves the case—also ending the process timeline. An agency is incentivized to take corrective action prior to submitting its report because it typically will not have to reimburse the protester's costs if the protest is sustained.[33]

In Figure 4.11 we show the distribution of the number of days it took to close all protest actions between FY 2008 and FY 2016. We measured the days to close as the number of days

Figure 4.10
GAO Bid Protest Timeline

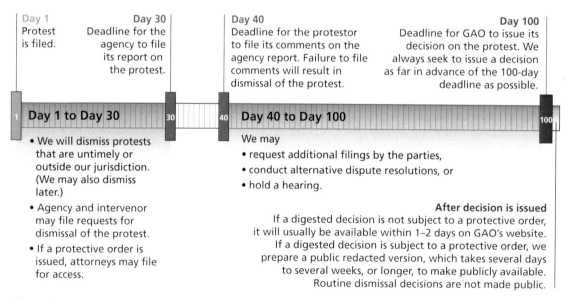

SOURCE: U.S. Government Accountability Office, "Bid Protests: Our Process," webpage, undated(d).
RAND RR2356-4.10

---

[32] See GAO, 2009a.

[33] GAO, 2009a.

Figure 4.11
Days to Close a Protest Action at GAO, FYs 2008–2016

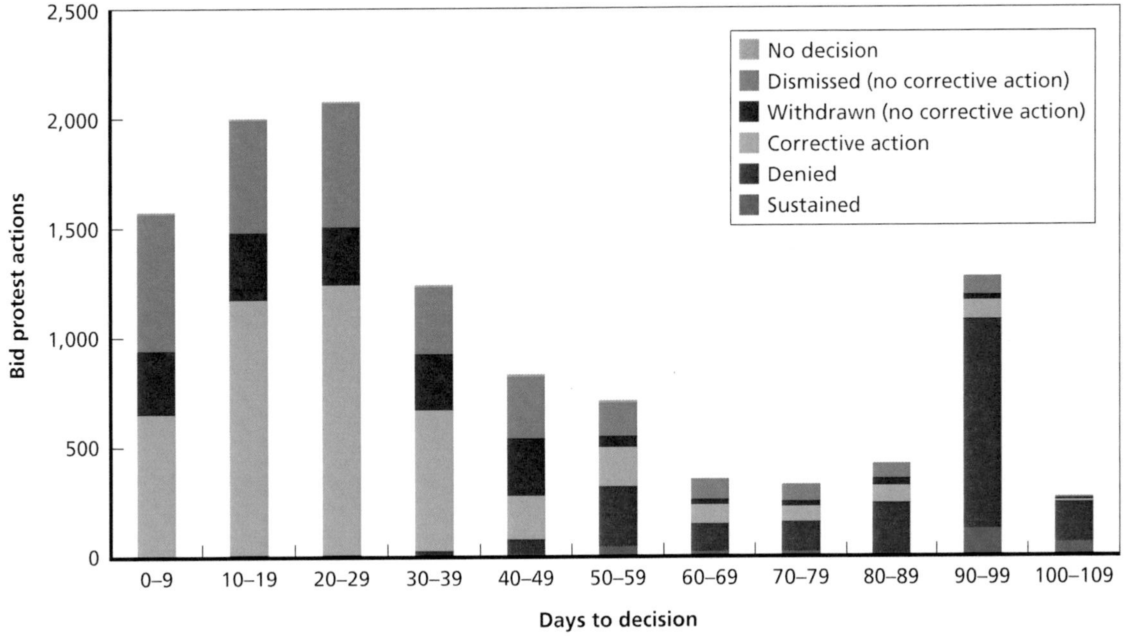

SOURCE: RAND analysis of GAO data.
NOTES: Excludes reconsiderations. All protest cases were resolved within the 100-day window.
The interval of 100–109 days includes only decisions that took 100 days.
RAND RR2356-4.11

between the date the protest was filed and when the protest was resolved (closed). We excluded reconsiderations from this timeline because they are not subject to the 100-day window.

The figure shows the distribution of the days to close as a function of how the case was resolved. There are two peaks to this distribution. The first peak is just under 30 days. (Not surprisingly, this corresponds to the due date for the administrative record.) Within this time frame, the cases are dismissed, withdrawn, or have corrective action.[34] Almost all the corrective action that occurs happens within 40 days, which corresponds to the agency report timeline. As stated earlier, the agency typically does not owe costs to the protester if it takes corrective action within the reporting deadline. Most dismissals happen early as well.[35] The next peak occurs between 90 and 100 days. Protests that survive the initial 40 days generally go to a formal GAO decision to sustain or deny them.

Figure 4.12 shows the cumulative percentage of cases resolved by days to close for three different populations of the protest data: all cases, ADR, and merit cases. The green line (all cases) shows that 50 percent of the protest actions are resolved within 30 days. Fully 70 percent of cases are resolved within 60 days. However, the blue line (merit cases) shows that if a case goes to decision it takes nearly the full window of 90–100 days.

---

[34] Note that cases that have corrective action are noted in the GAO protest docket as formally withdrawn or dismissed as academic. We have separated those corrective action cases out in Figure 4.11.

[35] For example, for not being timely, being legally insufficient, or if GAO lacks jurisdiction.

Figure 4.12
Days to Close Merit Cases, Cases Using ADR, and All Cases at GAO

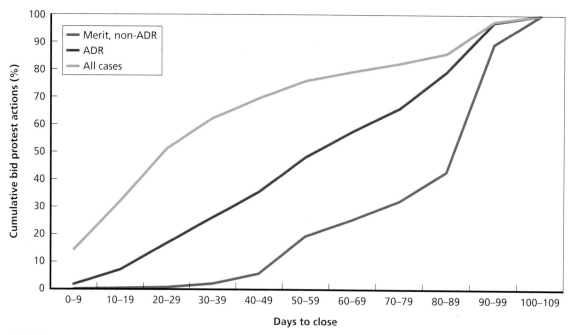

SOURCE: RAND analysis of GAO data.
NOTE: Excludes reconsiderations.
RAND *RR2356-4.12*

This time frame has important implications for decisionmakers. There had been some debate as to whether the GAO timeline should be shortened from 100 days to 65 days.[36] While a large number of the protest actions are resolved within the 65-day window, those requiring decisions by GAO are not. These cases are typically more complex and as such are not simply resolved. A concern would be that shortening the window for GAO to close a protest might not leave adequate time to develop these more complex decisions. Alternatively, it may entail additional resources at GAO to resolve cases. These trade-offs and limitations should be fully considered before reducing GAO's decision window for protests.

Quantitative Observations Specific to GAO

From the data we reviewed, we highlight several GAO-specific observations:

- The stability of the effectiveness rate over time—despite the increases in overall protest actions—suggests that firms are not more or less likely to protest without merit.
- Small-business protests are less likely to be effective and more likely to be dismissed for legal insufficiency.
- Cases in which legal counsel is required (i.e., a protective order was issued by GAO) have higher effectiveness and sustained rates.

---

[36] Yang, 2017.

- Protest filing peaks at the end of the fiscal year.
- Protests on task-order solicitations have a slightly higher effectiveness rate.
- There are measurable differences between the defense agencies, but DoD has a slightly lower overall effectiveness rate than non-DoD agencies.
- The largest DoD contractors have slightly higher sustained and effectiveness rates, but these differences are diminishing with time.
- DoD uses stay overrides infrequently.
- The number of protesters and protest actions tends to grow with a contract's value.

We also identified several observations common to both GAO and COFC. Those observations are presented at the end of Chapter Five.

Quantitative Analysis of DoD Bid Protest Activity Since CY 2008 at COFC

In this chapter, we address the RAND team's quantitative analysis of COFC bid protest experiences. We used data compiled and provided by the court on activities between 2008 and mid-2017.[1]

COFC Data Characterization

COFC provided the RAND team with details of 475 cases with filing dates from January 2008 through May 2017 in which a DoD agency was involved. These records were compiled by the court clerk's office from its case docket system. A number of details were provided for each case, including the protester, DoD agency, procurement value, case outcome, number of days until administrative record was filed, the number of days the case was pending, the number of intervening defendants, whether the case summary has a prior mention of GAO,[2] and whether the protester was a small business. We will explore the COFC protest activity at the case level. The court has what are termed "related cases," but it was not possible to link individual cases to other related ones. Regardless, case decisions are independent and, as such, we treated them that way.

One important fact regarding the court's bid protest cases is that protests represent a fraction of the court's overall caseload. Figure 5.1 shows the fraction of cases at the court that were bid protests between CYs 2008 and 2016. This fraction has been less than 20 percent, on average, but varies with calendar year.

COFC Time Trends

The overall number of protest cases at the court has been steadily rising. Figure 5.2 shows the number of bid protest cases for DoD and non-DoD agencies in CYs 2008–2016. DoD cases represented approximately half of total protest cases. While the DoD trend is not statistically significant in isolation, the non-DoD trend and overall combined trends upward are significant.

---

[1] The GAO data presented in Chapter Four covered FY 2008 through the end of FY 2016. The COFC data discussed in this chapter span the start of CY 2008 through mid-2017.

[2] We attempted to work with the clerk's office to see if we could directly link GAO and COFC protests. Unfortunately, this was not possible due to the lack of specific identifiers that would provide a definitive way to link cases. As a proxy, we used references to GAO in the case details.

Figure 5.1
Bid Protests as a Fraction of COFC's Overall Caseload

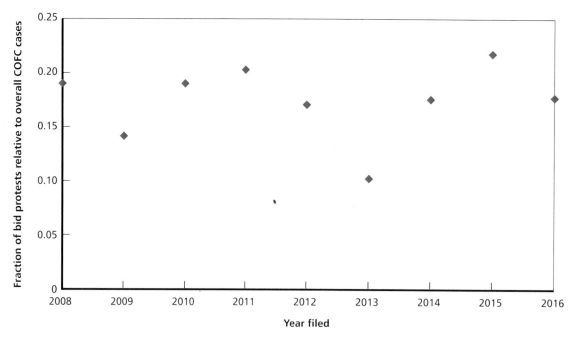

SOURCE: RAND analysis of COFC data.
RAND *RR2356-5.1*

Figure 5.2
Protest Cases at COFC, DoD Versus Non-DoD, CYs 2008–2016

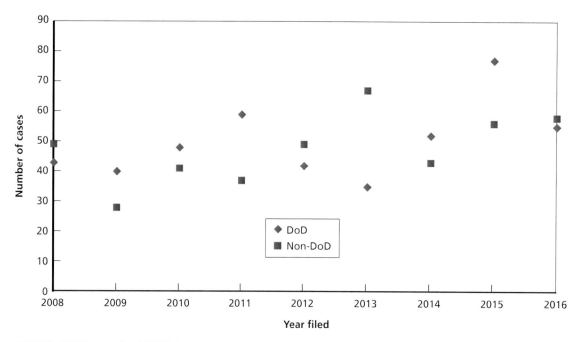

SOURCE: RAND analysis of COFC data.
RAND *RR2356-5.2*

Figure 5.3 shows the percentage of contracts protested and the number of procurements protested per billion dollars of spending from FY 2009 through FY 2016.[3] The time trend for the percentage of contracts protested is not statistically significant, whereas the trend per billion dollars of spending is marginally significant (positive). Similar to the GAO data, the protest rates per contract and billion dollars of spending are low (less than 0.025 percent and 0.3 protests per billion dollars of spending, respectively). These low values are consistent with the prior observation that very few procurements are protested at COFC.

COFC Protests by DoD Agency

As we saw in the GAO protest data, there are differences in protest activity between the DoD agencies at COFC. We used the same five groupings of agencies as we did for GAO. Figure 5.4 shows the number of protest actions by agency and calendar year. Similar to GAO protest activity, the Army has more protest activity than the other services at COFC. Unlike the GAO data, there appear to be no significant trends with time, and the year-to-year protest numbers vary. Table 5.1 displays the percentage of COFC cases relative to spending and contracts for the five agency groupings from FY 2009 through FY 2016. However, the relative

Figure 5.3
DoD Percentage of Procurements Protested and Protests per Billion Dollars at COFC, FYs 2009–2016

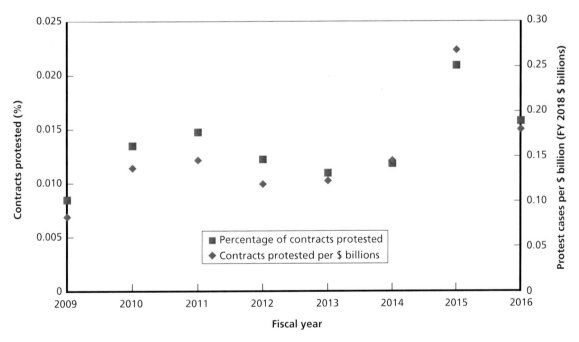

SOURCE: RAND analysis of COFC and FPDS-NG data.
NOTE: Complete data for FY 2008 were not available.
RAND RR2356-5.3

---

[3] For our COFC data analysis, we generally used calendar years because the data window ended and began on the calendar year. For comparison with contract numbers and spending, we converted to fiscal year so that the data were aligned. Because FY 2008 was incomplete, we omitted it from analyses involving fiscal years.

Figure 5.4
COFC Bid Protest Cases, by Agency and Year Filed, CYs 2008–2016

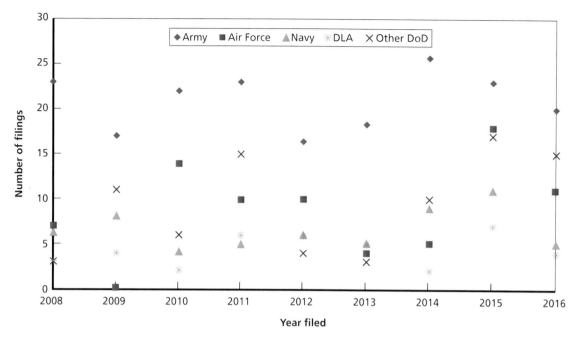

SOURCE: RAND analysis of COFC data.

RAND RR2356-5.4

Table 5.1
Percentage of Protests, Spending, and Contracts by DoD Agency at COFC, FYs 2009–2016

Agency	% of DoD Protest Cases	% of DoD Contract $	% of DoD Contracts
Army	41	34	25
Air Force	17	19	9
Navy	13	28	19
DLA	9	11	44
Other DoD	19	9	3

SOURCE: RAND analysis of COFC and FPDS-NG data.

NOTE: Columns may not sum to 100% due to rounding.

level of contract spending does not correlate well with the number of protest cases (as was the case for the GAO data). Still, the Army had relatively more protests compared with its spending or number of contracts; the Navy and DLA were relatively lower compared with their relative share of contract spending. Other DoD agencies had a much greater proportion of cases at COFC than at GAO.

Firms with Largest Funds Awarded

The top 11 firms (by FY 2016 revenue) have filed few cases at COFC. In all, those firms filed just ten protest cases between CY 2008 and CY 2016. Seven were filed by one firm, L3 Technologies (different divisions), with three of those cases being related. Again, the use of protest by the largest providers to DoD is infrequent, suggesting that protests at COFC are not part of standard business practice for these firms.

Pattern of Protest Filings at COFC

Unlike GAO, there is no pattern to the filing of cases at COFC across the year. Figure 5.5 displays the cumulative number of DoD protest cases filed by calendar month between CY 2008 and CY 2016. There is no statistically significant pattern either by month or quarter. Moreover, there is significant variability year to year such that the peak month is not consistent. Interestingly, September is the lowest month (the end of the fiscal year) unlike at GAO, where it was the second highest.

Protest Characteristics

COFC tracks a number of characteristics associated with procurements, such as approximate acquisition value, whether the protester is a small business, and whether the protest is appealed

Figure 5.5
DoD Bid Protest Filings at COFC, by Month, CYs 2008–2016

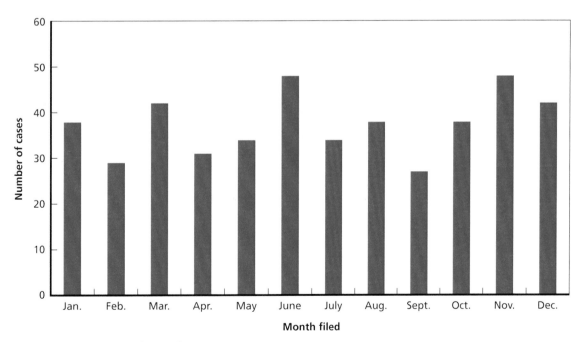

SOURCE: RAND analysis of COFC data.
RAND RR2356-5.5

to the Circuit Court of Appeals. In Table 5.2, we summarize these characteristics by case. Similar to GAO, the majority of protesters at COFC were self-identified small businesses. Again, this dominance of small businesses at both GAO and COFC suggests that policy changes to improve the bid protest system should also consider small-business issues. A nontrivial number of protests were for contracts less than $0.1 million in value—3.5 percent. Similarly, a policy area to further explore is whether adjudicating protests at procurement values this low is efficient. On average, the value of protested procurements is approximately $1.1 million (FY 2018 dollars). However, this average is a bit misleading because the distribution of procurement values is very skewed (very close to a log-normal distribution).

Figure 5.6 compares the protested procurements' acquisition values as reported by the protester. Cases at COFC tend to be higher in value than at GAO. The shift to higher procurement values is, perhaps, not surprising, as the costs to file protests at COFC presumably are higher due to the requirement to be represented by legal counsel (unlike at GAO, where representation is optional). GAO was set up to be an "inexpensive and expeditious forum for the resolution of bid protests."[4] However, during our discussions, some disputed whether there was a real difference in the costs between the two venues.

To help us determine whether protest cases had appeared before GAO prior to coming to the court, the COFC clerk's office provided a list of cases that included a reference to GAO in their case summary. More than 50 percent did, on average. While this is an inexact measure and care must be exercised in its interpretation, it nonetheless suggests—but does not prove—that a large fraction of cases at COFC were filed previously at GAO, where the protester did

Table 5.2
DoD Bid Protest Characteristics at COFC, CYs 2008–2017

Characteristic	All Cases
Observations	475
Average number of intervenors[a]	0.6
From small businesses[b]	58%
Value under $0.1 million[b]	3.5%
Average value (FY 2018 $)[b]	$1.1 million
Reference to GAO	52%
Percentage of cases appealed to U.S. Circuit Court	12%

SOURCE: RAND analysis of COFC data.

NOTES: The table includes cases from CY 2017, but the year was incomplete at the time the data were collected and therefore we were not able to explore time trends. However, we include these observations in the sample averages.

[a] Intervenors are firms that enter protests to protect their status as an awardee or potential awardee.

[b] These values are self-reported by the protester.

---

[4] See U.S. Government Accountability Office, "Bid Protests: Search Protests," webpage, undated(c).

Figure 5.6
Comparison of Protest Acquisition Values Between GAO and COFC

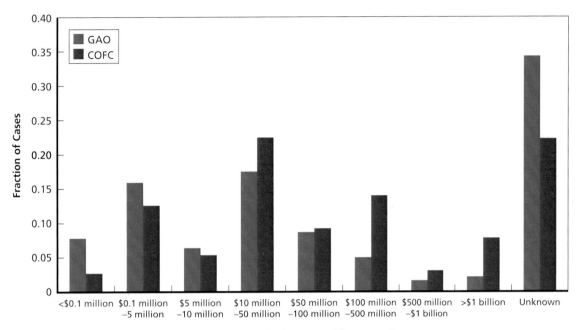

SOURCE: RAND analysis of GAO and COFC data.
RAND *RR2356-5.6*

not achieve the outcome it wanted.[5] Figure 5.7 shows how this percentage of cases referencing GAO is growing with time. The recent level of around 70 percent is similar to the anecdotal value we heard from one agency about the number of cases at COFC with a prior history at GAO. The increase with time could suggest that companies may be more willing to protest at COFC if they lose at GAO—possibly linked to declining procurement funding.[6] It could also reflect dissatisfaction with corrective action. Unfortunately, we were not able to identify cases for which corrective action was being protested at COFC.

DoD Bid Protest Outcomes at COFC

Table 5.3 summarizes the protest outcome measures for DoD protests at COFC. Note that we did not have as many outcome measures as we did for GAO protests. For example, the COFC case records did not indicate when corrective action took place, just that the case was dismissed.

The sustained rate appears to be significantly declining with time. Figure 5.8 shows trends from FY 2008 through FY 2016. The rate has been cut nearly in half, but the magnitude is

---

[5]   Others have observed rates at this level as well. See Saunders and Butler, 2010.

[6]   We will see in the next section that these protest cases (those with a reference to GAO) do not have different sustained rates, which suggests that companies are similarly selective in the cases they bring to COFC after going to GAO as they are with other cases. Otherwise, one might observe a lower relative sustained rate. However, it is true that the sustained rate is broadly declining with time at COFC and the GAO-related cases are following a similar pattern. Further research into how frequently bid protests appear between the two venues and whether there are ways to reduce multiple protests for the same procurement would be helpful to decisionmakers.

Figure 5.7
Fraction of COFC Cases That Referenced GAO, CYs 2008–2016

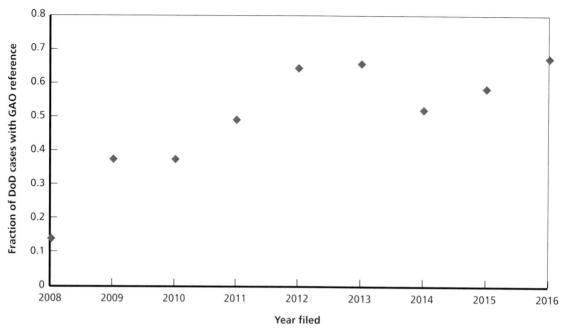

SOURCE: RAND analysis of COFC data.

RAND *RR2356-5.7*

Table 5.3
DoD Bid Protest Outcome Measures at COFC, CYs 2008–2016

Outcome Measures	All Cases
Observations	459[a]
Sustained rate	9%
Appeals rate	12%
Percentage of appealed cases sustained	20%

SOURCE: RAND analysis of COFC data.

[a] The number of observations for cases with decisions is lower than in Table 4.3 because some cases were not yet decided.

somewhat uncertain due to year-to-year variability.[7] The reasons for the decline in the sustained rate at COFC are unclear. This decline, coupled with the previous observation that the number of cases that reference a GAO protest is increasing, could suggest that firms are more likely to bring protests to the court that lose at GAO (or do not achieve the desired outcome). However, as we discuss in the next section, GAO protests have similar sustained rates. This decline is a broader trend at COFC and could suggest that, in general, protesters are being less selective

---

[7] The number of sustained cases in any year ranges from one to nine. Thus, the sustained rate reflects a few, discrete observations. One or two cases decided differently would have a substantial effect on the rate. Therefore, any trend must be viewed cautiously. However, there is a statistically significant trend downward over time.

Figure 5.8
Sustained Rate by Year for DoD Protests at COFC, CYs 2008–2016

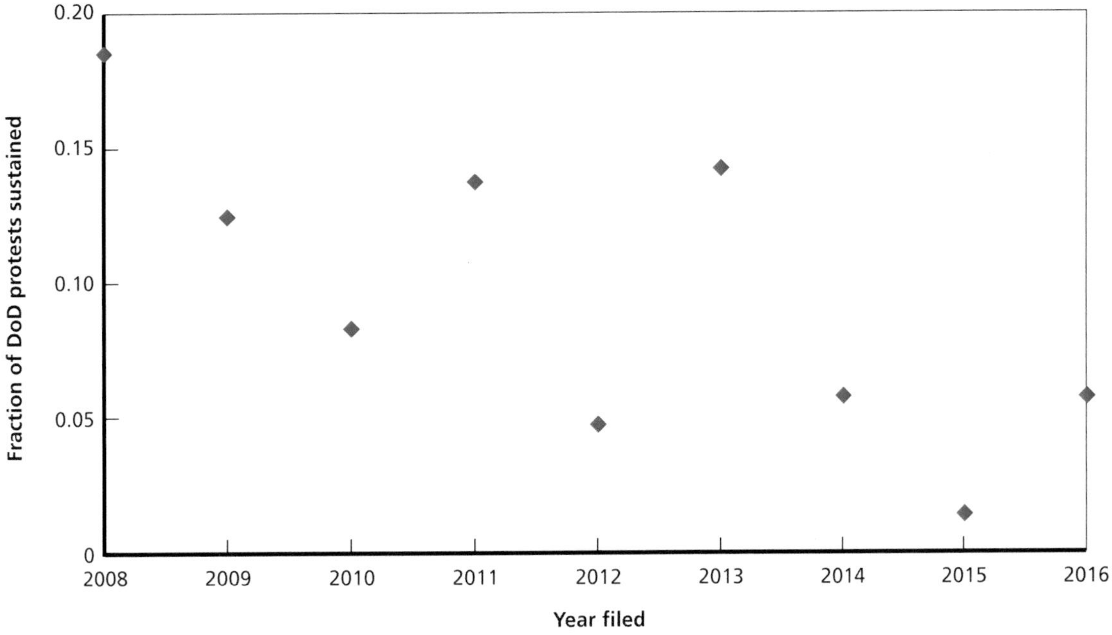

SOURCE: RAND analysis of COFC data.
RAND *RR2356-5.8*

in the cases they bring to COFC. Further research in this area would provide decisionmakers with a greater understanding of the reasons for this decline and possible improvement actions.

As discussed earlier, we explored the correlation of the sustained rates to address themes raised in Section 885 of the FY 2017 NDAA. Table 5.4 shows the correlation of the characteristics with the sustained rate at COFC.[8] As before, a "+" indicates a positive correlation (a higher rate), a "−" indicates a negative correlation (lower rate), and a "0" indicates no statistically meaningful correlation. Two repeated symbols indicate the correlation was strong (a relatively large coefficient).

Aside from the time trend of decreasing sustained rate and the higher rate of appealed cases that are sustained, there are no other observed correlations. There is no difference between the agencies, unlike at GAO, where there were very different rates. Also unlike at GAO, small-business protests did not differ. This result suggests that when small businesses are forced to use legal counsel, their protest sustained rates are similar to those of larger firms. Cases that had a reference to a GAO protest had the same sustained rate as those that did not. If this GAO reference is representative of protests that had lost earlier before GAO, then it does not appear that such cases are more likely to lose at COFC. There is also no difference in outcome by procurement amount or the number of intervening defendants.

COFC Bid Protest Timelines

The protest process at COFC begins when a protester files a complaint. The subsequent timing of the case, in part, revolves around the filing of the administrative record by the defending

---

[8]   Logistic regression details are shown in Appendix A.

Table 5.4
Correlation of Protest Characteristics with Protest Outcomes at COFC

Characteristic	Sustained Rate
Army	0
Air Force	0
Navy	0
DLA	0
Other DoD	0
Small business	0
Year filed	– –
GAO reference	0
Appealed case	+ +
Procurement amount	0
Number of intervenors	0

SOURCE: RAND analysis of COFC data.

NOTES: + + = strong, positive correlation, 0 = no correlation, – – = strong, negative correlation. See Appendix A for details on the logit analysis and how we characterized "strong" correlations. Small-business status was self-reported by the protester.

government agency. Before the administrative filing, a number of actions occur. The court randomly assigns a judge to the case. Typically, within 24–48 hours of filing, the judge will hold a scheduling conference to set timing for the protest case and to determine the status of the procurement. This initial hearing generally occurs once the Department of Justice attorney representing the government is selected. The court also determines early in the case whether the government will voluntarily maintain the status quo on the procurement until the case is resolved—a course of action that typically occurs. In circumstances in which the government does not agree, the protester may request a temporary restraining order hearing from the court. However, the criteria for the issuance of such an order are quite high, and the protester may have to file a bond if successful.[9] Protective orders are also issued at this early, pre–administrative record stage.

Once the administrative record is filed, the parties respond and motions are filed by the protester, government, or interested parties (intervenors) for various outcomes (e.g., dismissal of the case, discovery, judgment on the administrative record). After all motions and responses have been filed, the court holds oral arguments where parties present their views. After oral arguments are complete, the court rules on the case in a written decision.[10]

The majority of cases have their administrative record filed within 20 days. Figure 5.9 shows the distribution of the days until the administrative record is filed. The average was 37 days and the median was 17 days. There were no differences between the DoD agencies.

---

[9] See Schaengold, Guiffre, and Gill, 2009.

[10] In unusual circumstances, such as an urgent need, the court may rule from the bench without a written decision.

Figure 5.9
Number of Days to File an Administrative Record with COFC, CYs 2008–2017

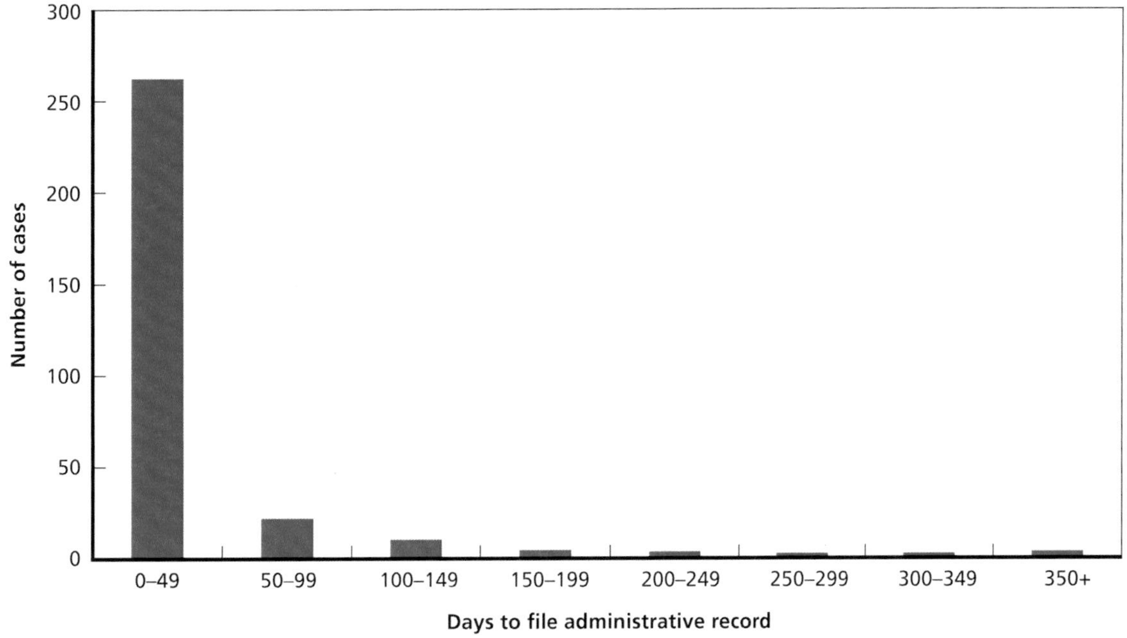

Days to file administrative record

SOURCE: RAND analysis of COFC data.
NOTE: Data cover the period through May 2017.
RAND RR2356-5.9

Note the long tail to the distribution of longer times (more than a year in some cases). This distribution is driven by the time measured from when the case is filed to when the *final* administrative record is filed. The administrative record may be supplemented or amended during the bid protest proceedings. Thus, the longer times are generally associated with cases for which the record was changed in some way. For example, a ruling for the case could include an update to the administrative record.

Figure 5.10 shows the distribution of the number of days between the start of cases and when a decision is issued at COFC. Seventy-five percent of the cases were resolved within 150 days. The average was 133 days and the median was 87 days. However, as with the days to file the administrative record, some cases take considerably longer. Interestingly, there appears to be no trend in terms of calendar year; that is, cases in a given calendar year did not take more or less time. This lack of a trend has persisted despite the recent decline in the number of active judges (see Appendix A, Figure A.10).

It is difficult to interpret the reasons for the longer decision times. Based on our discussions with court officials and our examination of long-duration proceedings that they provided as exemplars, cases may be left open after an initial decision for several reasons, including circumstances in which

- administrative record filings or updates are not produced in timely fashion
- additional follow-up actions are filed
- appeals are filed
- cases are held for corrective action determinations.

Figure 5.10
Number of Days to Close Cases with COFC, CYs 2008–2017

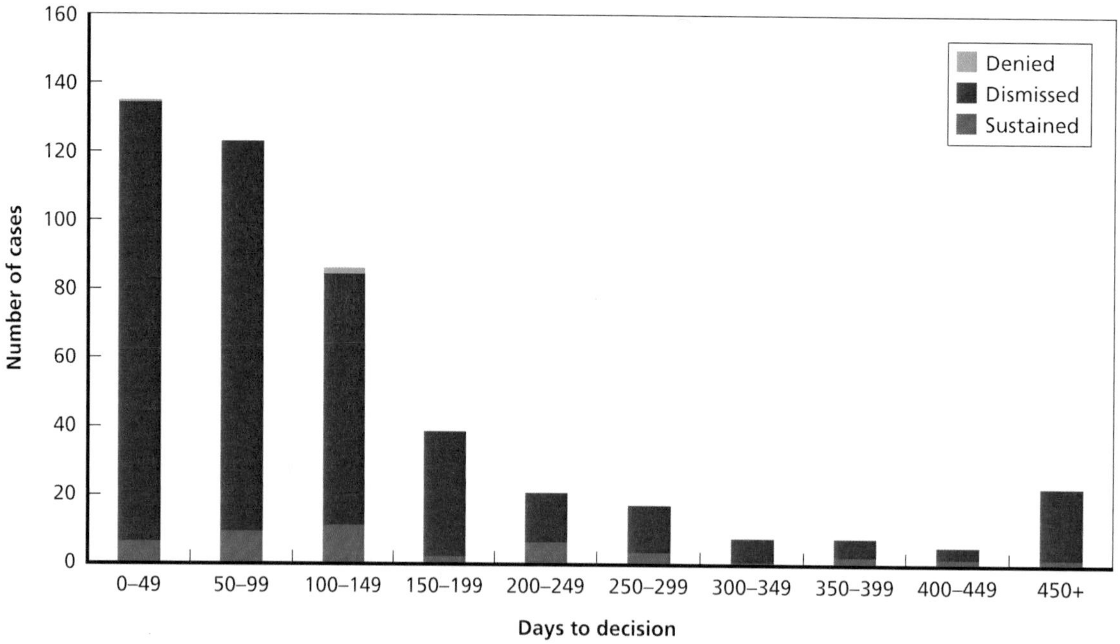

SOURCE: RAND analysis of COFC data.
NOTE: Data cover the period through May 2017.
RAND RR2356-5.10

Note that if a case is appealed to the U.S. Circuit Court, it is left open and time continues to accrue. Only when the appeals ruling is issued is a case closed. Therefore, some longer cases go through a judicial process twice. A case may be held open until an agency determines its approach to corrective action and the court approves.

Quantitative Observations Specific to COFC

From the data we reviewed, we highlight several COFC-specific observations:

- The sustained rate at COFC is declining with time as the number of cases increases. These trends suggest that firms may be more willing to file protests with COFC.
- There are no differences in sustained rates between DoD agencies or for small businesses relative to larger ones.
- The appeals rate is declining over time.
- Data and discussions suggest that the number of COFC cases that previously appeared at GAO may be increasing, but this potential trend needs further research.

Quantitative Observations Common to GAO and COFC

The analyses in Chapters Four and Five produced several observations common to both GAO and COFC:

- While our statistical modeling indicates differences between characteristics of the protest cases, these models cannot be used to reliably predict case outcomes.
- The overall level of bid protest activity has been increasing at both GAO and COFC since 2008.
- Bid protests by small-business plaintiffs represent the majority of protests at both venues.
- At both venues, in a nontrivial number of cases (approximately 4–8 percent), the contract value is less than $0.1 million (then-year dollars, as reported by the protester).
- There are differences between DoD agencies in terms of the number of cases filed. Specifically, the Army has the highest number of cases, year-on-year at both venues. This is partly explained by its share of contract expenditures.
- Trends differ between GAO and COFC, suggesting that any changes to the protest system should be tailored to the venue. For example, COFC's sustained rate is declining whereas at GAO it is holding steady (and potentially increasing).

CHAPTER SIX

Supplemental Data and Analysis

In the previous two chapters, we explored recent bid protest experiences at GAO and COFC based on their formal records. Those chapters identified important trends and differences in the bid protest data. There were, however, several issues raised in Section 885 of the FY 2017 NDAA that these official data were not able to support, such as incumbent protests and bridge contract issues.

In this chapter, we attempt to address some of those unanswered issues by supplementing the existing data with data from Deltek's GovWin database.[1] GovWin is a commercial website that tracks many government contracts from solicitation release through award, as well as various contract modifications. Deltek compiles this information from various sources, including public repositories, such as FedBizOpps, and Freedom of Information Act requests.[2]

In this chapter, we focus on a subset of GAO protest data for which we could match records between GovWin and GAO through solicitation numbers.[3] Given that much of the matching occurred manually, one record at a time, the process was very time-consuming. Therefore, we matched as many protest actions as we could that were filed in FYs 2015 and 2016 (the most recent two years). We refer to these supplemental data as the *subset* data. *Importantly, the results in this chapter should be interpreted with caution. The sample was not random and may not be representative of broader protest activity.* For example, as we shall see, the subset is biased toward larger contract values. This analysis focused on four additional areas in which it was feasible to collect information:

- incumbent protests
- bridge contracts and contract extensions related to a protest
- protests related to support services
- delays due to bid protests.

We begin with a brief characterization of this subset of data, after which we discuss each of the topics above in sequence.

---

[1] For more information, see Deltek, "GovWinIQ: Grow Your Business and Bottom Line—Overview," webpage, undated.

[2] FedBizOpps is the U.S. government's website where solicitations for all agencies are posted. For more details, see question 19 at Federal Business Opportunities, "Frequently Asked Questions," webpage, undated.

[3] It was not possible to do a similar matching with the COFC data as it did not contain solicitation number information.

Data Characterization

In all, we matched and supplemented 476 out of 2,916 protest actions for those protests filed in FYs 2015 and 2016 from the original GAO data described in Chapter Four.[4] These subset data included 249 out of 1,841 individual procurements. As noted earlier, the subset is not representative of the full GAO protest data. In Table 6.1, we show the number of protest actions and procurements by DoD agency groupings for the two fiscal years. Whether looking at the percentages by protest action or procurement, the Army, Air Force, and Navy are slightly overrepresented, whereas DLA and other DoD agencies are underrepresented.

Similarly, the distribution of acquisition values was different for the subset data. Figure 6.1 shows the distribution of the reported acquisition value at the procurement level for the subset data. The data are presented at the procurement level rather than at the protest action level. Notice that the subset data tend toward larger procurement values.

Table 6.2 summarizes some key statistics from the subset data. Again, the sample has fewer protests below $100 million compared with the values in Table 4.1 in Chapter Four. The sustained and effectiveness rates are also higher, consistent with the overall shift in the distribution of acquisition values. The percentages of small business and pre-award protests are similar.

Incumbent Protests

We also used data from the GovWin database to identify whether a protester was an incumbent. In the "Description" tab, GovWin tracks incumbents and prior solicitation numbers. Using this information, we evaluated protesters' incumbent status, reporting it as *no* (not an incumbent protester), *yes* (an incumbent protester), *new*, or *unknown*. *New* indicated that the contracting office considered the requirement a new effort with no incumbents. *Unknown* was used when the incumbent information was not released or when we could not determine the

Table 6.1
Percentage of Protest Actions and Procurements by DoD Agency for Subset Data, FYs 2015–2016

Agency	% of DoD Protest Actions		% of DoD Protested Procurements	
	Subset Data	Complete Data	Subset Data	Complete Data
Army	41	36	41	34
Air Force	22	20	22	20
Navy	29	20	29	21
DLA	3	15	3	18
Other DoD	6	9	5	7

SOURCE: RAND analysis of GAO data.
NOTE: Columns may not sum to 100% due to rounding.

---

[4] By *supplemented*, we mean that some additional protest/procurement characteristic was added. It does not mean, however, that all additional characteristics were determined for all observations. For some procurements, we had only partial additional information.

Figure 6.1
Comparison of Acquisition Value by Procurement, FYs 2015–2016

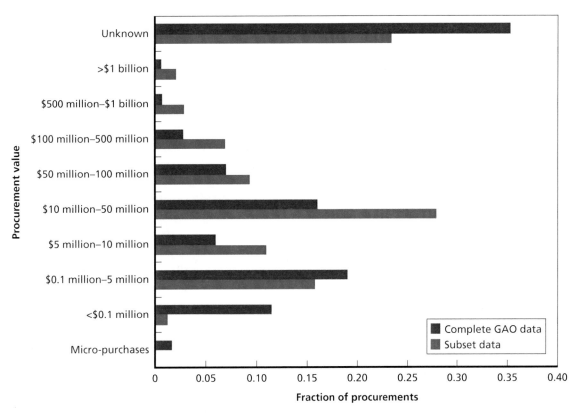

SOURCE: RAND analysis of GAO data.
RAND *RR2356-6.1*

Table 6.2
DoD Bid Protest Characteristics for Subset Data, FYs 2015–2016

Characteristic	All Actions	Weighted by Procurement
Observations	476	249
Average number of protesters	1	1.3
From small businesses[a]	50%	52%
Value under $0.1 million[a]	1.2%	1.2%
Task-order protests	18%	14%
Pre-award protest	28%	28%
Sustained rate	5.3%	3.2%
Effectiveness rate	44%	37%

SOURCE: RAND analysis of GAO data.
[a] Values were self-reported by the protester.

protester's status from the record. In Table 6.3, we summarize the incumbent bid protest statistics for the full subset data and task orders only. Approximately one-quarter of protest actions were associated with an incumbent. This fraction was nearly double for task-order protests, however. This difference is statistically significant and suggests—but does not prove—that incumbents are more likely to protest task orders when it may be to their economic advantage if they get a bridge contract during the CICA stay.

It may be, however, that incumbents have good reasons to protest task orders. In Table 6.4, we show the effectiveness rate versus whether the protester was an incumbent and whether the protests concerned a task order.[5] Note that when the incumbent protests a task order, the effectiveness rate is approximately 70 percent, which is much higher than average and statistically significant. Thus, while incumbents may protest task orders more frequently, incumbents are also much more likely to be successful.

Section 885 also requested an examination of incumbent protests over $100 million. Limiting the subset data to acquisition values greater than $100 million resulted in only 82 observed protest actions. This sample of protests over $100 million has a higher effectiveness rate of approximately 65 percent compared with procurements of lower value,[6] but there are no meaningful differences between task orders and incumbent populations for protests with values greater than $100 million.

Table 6.3
Incumbent Statistics from Subset Data, FYs 2015–2016

Incumbent Status	Full Subset (%)	Task Orders Only (%)
No	52	54
Yes	23	37
New	18	7
Unknown	7	2

SOURCE: RAND analysis of GAO and GovWin data.
NOTE: Averages are based on protest actions.

Table 6.4
Effectiveness Rate by Incumbent Status and
Task Order for Subset Data, FYs 2015–2016

Incumbent Status	Task Order (%)	
	No	Yes
No	41	42
Yes	47	71

SOURCE: RAND analysis of GAO and GovWin data.
NOTE: Averages are based on protest actions.

---

[5]   New and unknown values for incumbents are excluded.

[6]   This is a statistically significant difference.

Bridge Contracts and Contract Extensions

Section 885 requested an analysis of the prevalence of bridge contracts and contract extensions related to bid protests—particularly related to incumbents. Using contract adjustments that are recorded as part of the contract history in the GovWin database, we identified bridge contracts or contract extensions that occurred *after* a protest was filed and *before* work commenced. We denote these modifications (both bridge contracts and contract extensions) simply as *extensions*. One caution is that this measure of extensions is associative: We cannot be certain that the protest directly caused the contract modification. Given the timing of the extension, however, it seems likely that the protest was related to the extension.

In all, we identified 29 procurements that had an extension—out of 224 procurements for which we could determine whether there was an extension. Thus, roughly 13 percent of the procurements with a protest had some form of extension. There were no statistically meaningful differences related to whether the protester was an incumbent or whether the protest concerned a task order for the frequency of extensions. Additionally, there were no differences in the effectiveness or sustained rates when a bridge contract was issued. Not surprisingly, extensions were associated with longer delays. We cover this correlation later in this chapter when discussing delays.

Protests Related to Support Services

One interesting question posed in Section 885 of the FY 2017 NDAA is whether there are any differences in bid protests between product and service procurements. Using the description of the procurement data from GovWin, we identified protests that clearly described a procurement as support service–related (e.g., logistical support services, information technology support services, equipment maintenance and repair services). Out of 138 procurements that we could parse, 86 (or 62 percent) were related to support service procurements. This value seems high, given that task orders account for roughly 14 percent of the protest population in the subset data (see Table 6.2). Moreover, procurements identified as being for support services were highly correlated with task orders. It is possible that the descriptions were more complete for these service procurements. In any case, this topic deserves further research to more fully clarify the prevalence of these contracts in the full protest population.

Delays

One key impact measure for bid protests requested by Section 885 was the resulting delays to procurements. We estimated delays due to protests by one of three methods, in order of preference:

- Adjustments to the timeline: For this method, we used the difference between the procurement timeline released just before the protest and the timeline released after a protest decision. The change in the expected award date or the expected commencement of performance determined the delay.

- Contract adjustments associated with the schedule (i.e., period-of-performance adjustments or lifts of stop-work orders): The duration of stop-work orders is a straightforward measure to determine and translate into a delay. The delay is the difference between the start date of performance and the date that the stop-work order is lifted. Many schedule modifications do not have a description (i.e., a stop-work order or period-of-performance adjustment), so an adjustment to the expiration date may occur due to some non-protest event. Thus, we only included adjustments for the purpose of calculating delays within a month of protest decisions.
- Duration of bridge contracts or incumbent contract extensions issued during the protest period: The duration of these extensions translates directly to the estimated delay.

This information was all part of the GovWin "timeline" and "contracts" data. The absence of an estimated schedule delay, or an estimate of "zero" delay, does not necessarily indicate that there was no delay. It means that we were unable to *discern* a delay from examining the records. For example, it could be that either no original timeline was released for the procurement or all delays happened without formal contract action. Some procurements had a gap between contract award and the start of the period of performance. If the protest was resolved within this gap, no delay would be reflected in the data. As with all measures in this chapter, some caution must be used in their interpretation.

There were 224 procurements from the subset data for which we evaluated whether there was a delay. Seventy-three of those procurements had a measurable delay. Again, we caution that these values *do not imply* that approximately one-third of procurements protested had delays, merely that we were *able* to measure delays for one-third of them. For only those procurements for which we could measure a delay, the average delay was 6.2 months and the median was five months. (Again, this is not the average delay on protested procurements but, rather, the average for procurements for which we could measure a delay.) The distribution is very skewed, as can be seen in the histogram in Figure 6.2.

There was no correlation between the protest outcome and the delay length. There was, however, a strong correlation with procurement value, with larger procurements tending to have longer delays, as shown in Figure 6.3.

Finally, delays were significantly correlated with bridge contracts and contract extensions. This is more of an associative than causal measure (e.g., bridge contracts result from delays and do not cause delays). There were no meaningful differences for delays between pre-award protests and post-award protests. Also, the average delay lengths were similar whether or not the cases were resolved through corrective action. However, cases that were sustained had a longer associated delay—an average of about ten months relative to an average of about six months for cases that were not sustained. While statistically different, there were only five sustained cases in the subset data. Thus, the difference can only be seen as suggestive.

Figure 6.2
Histogram of Delay Times for Procurements with a Measurable Delay

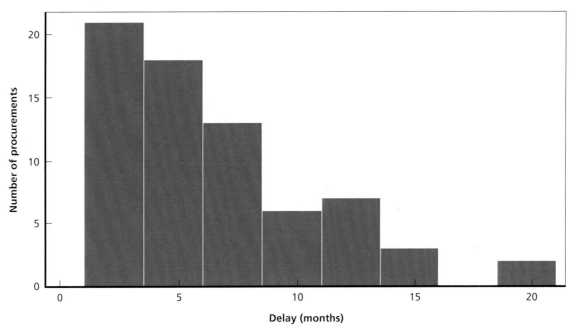

SOURCE: RAND analysis of GovWin data.
RAND *RR2356-6.2*

Figure 6.3
Delay Length Versus Procurement Value

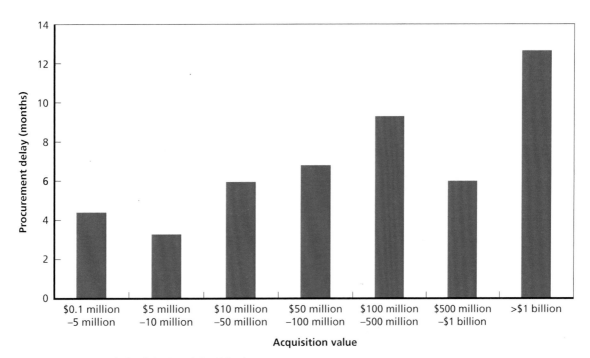

SOURCE: RAND analysis of GAO and GovWin data.
RAND *RR2356-6.3*

CHAPTER SEVEN

Recommendations

The observations and conclusions in previous chapters point to the following recommendations, which are intended to inform future changes to the DoD bid protest system. There is likely value in using the same or similar approaches across other U.S. government departments and agencies. In implementing these recommendations, there should be some consideration of costs and benefits, as some changes will require additional time or resources to implement.

Enhance the Quality of Post-Award Debriefings

A major concern from the private sector is the quality of post-award debriefings. The consensus among companies is that the quality and number of post-award debriefings vary significantly. The worst debriefings were characterized as being skimpy, adversarial, and evasive or as failing to provide required reasonable responses to relevant questions. In desperation, unsuccessful offerors may submit a bid protest to obtain government documents that delineate the rationale for the contract award. The bottom line is that, in most cases, too little information and evasive/adversarial debriefings will lead to a bid protest.

Our recommendation is to consider having DoD adopt a debriefing process similar to the U.S. Air Force's extended briefing process. The extended briefing process provides an unsuccessful offeror the opportunity to participate in an extended or enhanced post-award debriefing as an initiative to dissuade unsuccessful offerors from filing bid protests. Extended debriefings offer a transparent means whereby the Air Force provides an unsuccessful offeror's outside counsel with information that is not otherwise provided. By offering such documents—or even the entire agency record—before a GAO protest can be filed, an offeror's outside counsel can determine whether a bid protest is warranted and provide an opinion to the unsuccessful offeror about the fairness and impartiality of the evaluation and whether the award decision was rational. The Federal Aviation Administration has adopted a similar process and has seen results similar to the U.S. Air Force in that extended debriefings frequently result in an unsuccessful offeror's counsel dissuading the company from filing a bid protest or persuading it to withdraw a previously filed bid protest.[1]

We note that the U.S. Senate and House Armed Services Committees recently addressed bid protest reform for DoD acquisitions related to post-award bid protest debriefings. This has been documented in Section 818 of the FY 2018 NDAA, "Enhanced Post-Award Debrief-

---

[1] Interagency Alternative Dispute Resolution Working Group, "Electronic Guide to Federal Procurement ADR 2d: Ch.08—Improved Debriefings," webpage, undated(b).

ing Rights." The general expectation is that these provisions should substantially improve the quality and usefulness of post-award debriefings by providing additional transparency into the underlying DoD competition process.[2] Disappointed bidders will gain greater understanding of the evaluation and award process and can better analyze any potential protest grounds before filing a protest.[3]

Be Careful in Considering Any Potential Reduction to GAO's Decision Timeline

There had been some debate as to whether the GAO protest timeline should be shortened from 100 days to 65 days.[4] As we described in Chapter Four, most protest actions are resolved within 30 days. Fully 70 percent of cases are resolved within 60 days. Thus, most protest actions are resolved quickly. However, cases that go to decision (merit cases) typically take 90–100 days (which is the allotted time for GAO decisions). These cases are typically more complex and are not easily resolved. A concern is that shortening the timeline for GAO to close a protest might not leave enough time for it to develop these more complex decisions. Consequently, additional resources may be required at GAO to resolve cases. We recommend that these trade-offs and limitations be considered before reducing GAO's decision timeline for protests.

Moreover, bid protests are not filed at GAO uniformly throughout the year but, rather, peak near the end of the fiscal year. An advantage of the 100-day window is that it allows GAO to smooth the workload between peak and minimum filing periods (which are adjacent). The reduction of the timeline to 65 days, among other things, would provide less flexibility to GAO in managing its protest workload and may require additional staff to meet the peak filing demand. We recommend that the implications of this filing pattern also be considered when exploring reductions to GAO's decision timeline.

In all, the data suggest that it will not be easy to reduce the GAO timeline to 65 days for all protests.

Be Careful in Considering Any Restrictions on Task-Order Bid Protests at GAO

The threshold for DoD task-order protests has risen from $10 million to $25 million for defense agencies.[5] In fact, there was a brief period in which GAO did not have legal authority to hear protests on task orders because its previous authority had expired.[6] The increased threshold is perceived as a mechanism for reducing protests and their attendant costs and delays for DoD.

---

[2]   Richard B. Oliver and David B. Dixon, "Changes for Bid Protests in FY 2018 NDAA," Pillsbury Winthrop Shaw Pittman LLP, November 16, 2017.

[3]   Oliver and Dixon, 2017.

[4]   Yang, 2017.

[5]   Richard B. Oliver, Alexander B. Ginsberg, and Selena Brady, "Differing GAO Task Order Protest Thresholds," Pillsbury Winthrop Shaw Pittman LLP, January 3, 2017.

[6]   Jared Serbu, "Senate Backs Down from Attempt to Restrain Bid Protests, but Wants More Data," Federal News Radio, December 5, 2016. This authority has since been reinstated.

Such a change could reduce the number of task-order bid protests (which account for approximately 10 percent of the protest actions). However, as we observed in Chapter Four, task-order protests are generally *more* likely to be sustained or have corrective action compared with other types of protests. This result suggests that task-order protests fill an important role in improving the fairness of DoD procurements. We recommend caution in considering any further restrictions on task-order bid protests.

Consider Implementing an Expedited Process for Adjudicating Bid Protests of Procurements Valued Under $0.1 Million

A surprising result from the analysis (at least to the authors) was that roughly 8 percent of GAO protest actions and nearly 4 percent of protest cases at COFC concerned procurements with a declared value under $0.1 million. An interesting policy question is whether the costs to the government to adjudicate these protests exceed the value of the procurements themselves and thus are not cost-effective.[7] We recognize that cost-effectiveness may not be the most important criterion in adjudicating bid protests. Transparency and fairness of government spending might be the overriding consideration, with cost-effectiveness being secondary. However, we recommend that streamlined processes be considered for protests under $0.1 million (or some other suitably low value)—perhaps processes analogous to how traffic tickets are adjudicated in traffic court or how cases are adjudicated in small-claims court. A different approach would likely be needed for each venue. For example, COFC could "rule from the bench" on such smaller-value protests and not be required to generate written decisions. (This would limit the protester's ability to appeal, however.) One could require ADR for such small-value protests at GAO. Some discussion with each venue would be necessary to develop the most appropriate approach.

Another, but perhaps less desirable approach from a fairness perspective, would be to restrict such low-value procurement protests to the agency level. However, given the lack of readily available data on agency-level protests, reporting requirements would need to be implemented so that there is some confidence that the agencies can transparently, fairly, and objectively adjudicate them.

Consider Approaches to Reduce and Improve Protests from Small Businesses

More than half of the protests at GAO and COFC are from self-identified small businesses. While small businesses are awarded more than half of DoD contracts, such contracts represent only 15–20 percent of total contract dollars. This disparity raises another cost-effectiveness question: Should more than half of protest activity be focused on less than 20 percent of contract dollars?

The protest activity by small businesses suggests that any improvements to the bid protest system should also address small businesses. For example, the current "loser-pays" pilot program for DoD *excludes* businesses with annual revenues under $250 million (Section 827

---

[7] An analogy would be when a civil case is settled out of court because it is more cost-effective for both parties.

of the FY 2018 NDAA).[8] Furthermore, the fact that small businesses are generally less successful at GAO (but not at COFC) suggests that small businesses' reasons for protesting differ from larger businesses' reasons (a feature that was corroborated in our discussions). Sometimes, when debriefings are uninformative, small businesses lodge protests to gain understanding of why they lost a procurement. To the extent that this is the case, the changes to the debriefing process discussed earlier should help to eliminate some small-business protests.

Other changes related to small businesses could also be considered. Protests (by both large and small businesses) have a higher effectiveness rate at GAO when under a protective order. Small businesses are also more likely to have their cases dismissed for lack of jurisdiction or for being legally insufficient. These differences suggest that small businesses might benefit from better legal representation in filing protests at GAO.[9] One option would be to require all protests at GAO to be filed through legal counsel. However, this approach might be viewed as unfair, as small businesses might face more-significant economic barriers to filing than larger businesses. Another option would be to provide legal assistance to small businesses—perhaps through the Small Business Administration. Such advice might be useful if it is provided early enough that small businesses can determine whether they have valid cases, which could allow them to craft more-persuasive arguments and, possibly, reduce the number of dismissed protests.

Consider Collecting Additional Data and Making Other Changes to Bid Protest Records

Several potential changes to data collection and reporting could be considered to aid future research and decisionmaking.

Recommendations for DoD Agencies

At DoD agencies, several steps potentially could be taken, including the following:

- Collect agency-level protest data and provide an annual report (similar to GAO's) that summarizes protest activities (at a minimum, elements such as the number of cases filed, time to resolve a case, case outcomes, the reasons for sustained protests, accommodations made, procurement value, whether the protestor subsequently wins the contract, whether the protester is a small business, and if the protest subsequently appears at GAO or COFC).
- Collect and summarize reasons for corrective action.

---

[8]   Section 827 states that the "Secretary of Defense shall carry out a pilot program to determine the effectiveness of requiring contractors to reimburse the Department of Defense for costs incurred in processing covered protests." See, for example, Fred Konkel, "Bid Protests Decline in 2017," *Nextgov*, November 15, 2017.

[9]   It is interesting to note that there is no difference in the sustained rate for small businesses at COFC. This lack of a difference could be, in part, a result of being required to use legal counsel in the process.

Recommendations for GAO

At GAO, some additional information could help:

- Provide the protester's DUNS number and name to allow better tracking across protest actions.
- Collect and report the acquisition value (not a range) to aid in understanding trends by value. Note that GAO does report this information in its decisions, but it would be more useful to researchers as part of the reported docket information.

Recommendations for COFC

At COFC, several changes could be made to case records to facilitate research:[10]

- Provide the protester's DUNS number and name to allow better tracking across protest cases.
- Restore the small business field so that protests by small businesses can be easily identified.
- Record the reason for withdrawal (e.g., corrective action). Currently, it is too difficult to determine from the case docket when corrective action has occurred.
- While some cases at COFC might appear to take a long time, this is due in part to the way the timeline is tracked. For example, a case is held "open" during an appeal (even though a decision has been issued). COFC could track decision time (or dates) at three different points: time to first decision, time until case is closed, and the time for an appeal.
- One important question that we could not definitively answer was the rate at which protests appear at both GAO and COFC. We used a proxy for this measure, but our findings were far from exact. COFC could require protesters to report whether there has been a prior GAO protest or activity and possibly request the B number(s) to link records between venues.
- As observed earlier, the majority of cases at COFC are resolved within 90 days. However, some cases take substantially longer. COFC could consider recording the reasons for cases pending after more than 100 days as part of its records.

In summary, our observations, analyses, and conclusions indicate that the recommendations provided here can positively inform future changes to the bid protest system. As stated earlier, we feel that there is value in using the same or similar approaches across other U.S. government departments and agencies, though their costs and benefits must be taken into account, and decisionmakers should weigh the additional time and resources required to implement them.

---

[10] In our discussions, COFC personnel indicated that they have considered our suggestions to improve data collection and recordkeeping and are implementing changes.

Supplemental Analysis

In this appendix, we present supplemental information, tables, and figures that underlie some of the analyses in Chapter Four.

GAO Data

Figure A.1 shows time trends for DoD contracts in terms of both spending and number of contracts, as derived from FPDS-NG. There is a significant downward trend for each that runs counter to the trend in protest activity over the same period. Contract spending has declined by more than 30 percent, and the number of contracts has declined by more than 10 percent.

Figure A.2 shows the proportion of DoD bid protests at GAO and DoD contract spending as a fraction of total federal government contract spending. The two ratios track consistently, and both values have declined slightly between FY 2008 and FY 2016. Figure A.3 shows the fraction of DoD contracts and contract dollars going to small businesses.

Figure A.1
DoD Contract Spending and Number of Contracts, FYs 2008–2016

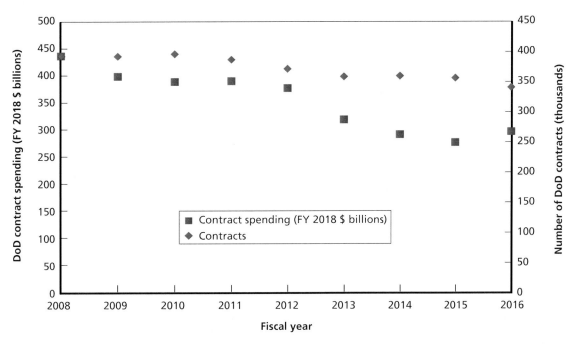

SOURCE: RAND analysis of FPDS-NG data.

Figure A.2
Relative DoD Spending and Protest Actions, FYs 2008–2016

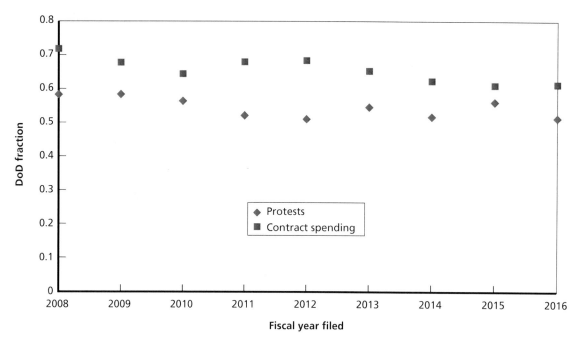

SOURCE: RAND analysis of GAO and FPDS-NG data.
RAND *RR2356-A.2*

Figure A.3
Fraction of Contracts and Contract Dollars Going to Small Businesses, FYs 2008–2016

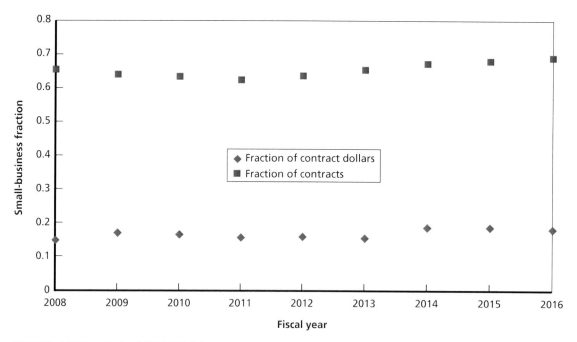

SOURCE: RAND analysis of FPDS-NG data.
RAND *RR2356-A.3*

Figure A.4 shows the distribution of protest actions and protested procurements by reported value (in ranges specified in the GAO data). Excluding unknown values, the modal range is $0.1 million to $5 million for procurements. The percentage of protests by procurement is lower than that for all protest actions at higher contract values. This is because larger-value procurement contracts are more likely to have multiple protest actions.

Figure A.5 shows the number of DoD protest actions by fiscal year, excluding task-order protests. The upward trend is still statistically significant. Figure A.6 shows the number of unique protesters for a procurement versus procurement value. As the procurement value increases, so does the number of protesters.

Figure A.7 shows the fraction of bid protest actions from small businesses by fiscal year. The increase is statistically significant. Figure A.8 shows the increase in the number of task-order protests over time. The trend is increasing, and it corresponds to GAO gaining jurisdiction over task-order protests in FY 2008.

Figure A.9 shows the time trend for the effectiveness and sustained rates of the top 11 firms. Both rates appear to be declining with time. The number of observations is relatively small, hence the variability in the year-to-year trend. These trends suggest that, while more successful, the rates for the top 11 firms have been approaching the overall sample average in recent years. This trend may also suggest that larger firms are becoming less selective in the cases they file.

Figure A.4
Reported Value of Protested Procurements

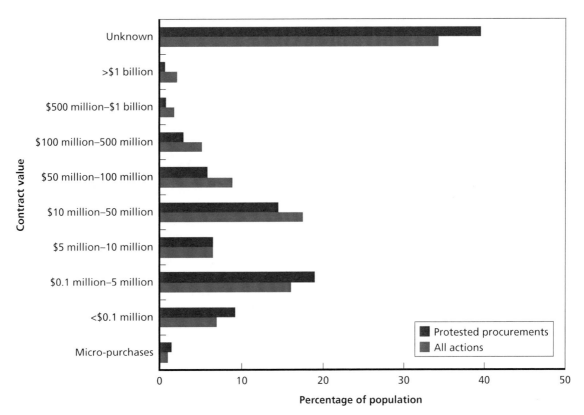

SOURCE: RAND analysis of GAO data.
RAND *RR2356-A.4*

Figure A.5
DoD Protest Actions Excluding Task-Order Protests, FYs 2008–2016

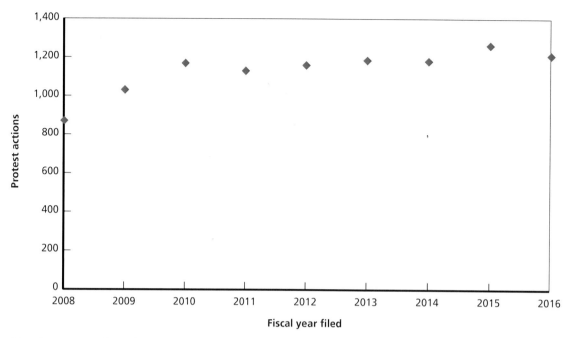

SOURCE: RAND analysis of GAO data.
RAND RR2356-A.5

Figure A.6
Number of Protesters by Procurement Value, FYs 2008–2016

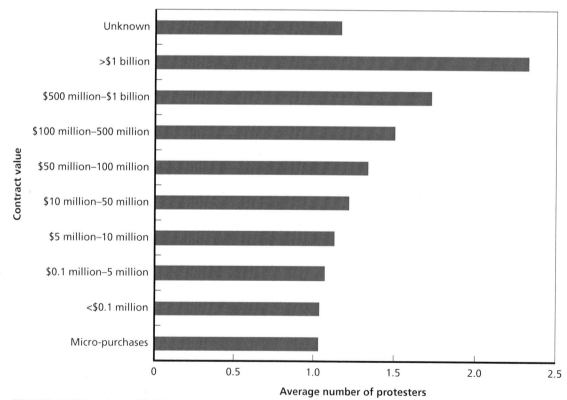

SOURCE: RAND analysis of GAO data.
RAND RR2356-A.6

Figure A.7
Fraction of Small-Business Protest Actions, FYs 2008–2016

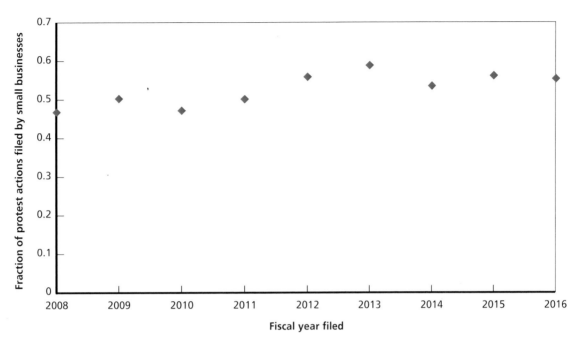

SOURCE: RAND analysis of GAO data.
RAND *RR2356-A.7*

Figure A.8
Number of Task-Order Protest Actions, FYs 2008–2016

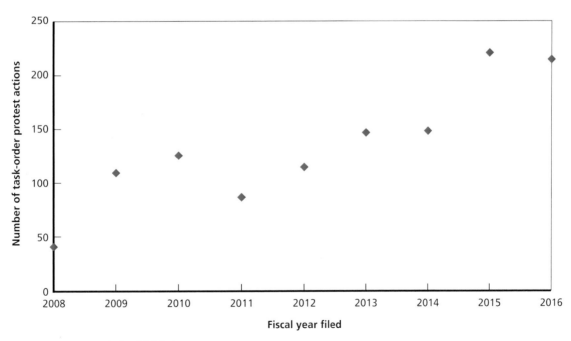

SOURCE: RAND analysis of GAO data.
RAND *RR2356-A.8*

Figure A.9
Effectiveness and Sustained Rates for Top 11 Firms at GAO, FYs 2008–2016

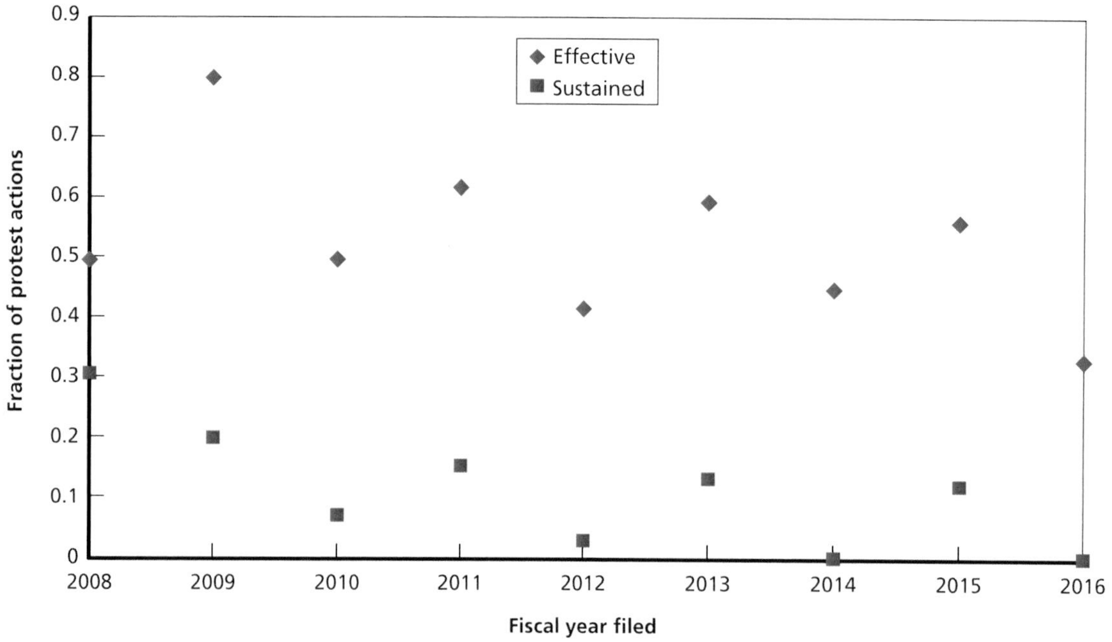

SOURCE: RAND analysis of GAO and FPDS-NG data.
RAND RR2356-A.9

Tables A.1 and A.2 show the logistic analysis diagnostic output for effectiveness and sustained rates at GAO, respectively. The odds ratio terms show the effect of the term on the relative odds of an outcome. For example, an odds ratio of 1.5 is interpreted as 50-percent more likely, all other things being equal.[1] In Tables 4.4 and 5.4, the symbols (e.g., +, −, 0) corresponded to the odds ratio magnitude. A "+ +" was used for odds ratios greater than 1.75, a "+" for odds ratios between 1.75 and 1.0, "0" for terms that were not significant, "− −" for odds ratios less than 0.4, and a "−" for odds ratios between 0.4 and 1.0.

COFC Data

Table A.3 shows the logistic regression for the sustained rate at COFC. Figure A.10 shows the change in the number of active judges at COFC. Figure A.11 displays the declining fraction of the cases appealed.

---

[1]   See StataCorp LLC, undated, for more details.

Table A.1
Logistic Analysis of Effectiveness Rate at GAO

. sw logistic Effective TaskOrder small Army AirForce Navy DLA Reconsideration rfp rfq Latvian Override Top11 preaward ADR NumProtesters InitialFiling if DoD==1&protest==1, pr(0.05)

                    begin with full model

p = 0.3955 >= 0.0500 removing Navy

Logistic regression			Number of observations		=	11,459
			LR chi^2(15)		=	725.37
			Prob > chi^2		=	0.0000
Log likelihood = −7393.3668			Pseudo R^2		=	0.0468

EffectiveProtest	Odds Ratio	Standard Error	z	P>|z|	[95% Confidence Interval]	
TaskOrder	1.20289	0.0768517	2.89	0.004	1.061312	1.363353
small	0.8231265	0.033623	−4.77	0.000	0.7597956	0.8917361
Army	1.14797	0.0557009	2.84	0.004	1.043828	1.262501
AirForce	1.140421	0.0684099	2.19	0.028	1.013922	1.282702
InitialFiling	1.269875	0.0582978	5.20	0.000	1.160603	1.389435
DLA	0.8363554	0.0579921	−2.58	0.010	0.7300781	0.9581034
Reconsideration	0.028245	0.0116799	−8.63	0.000	0.0125589	0.063523
rfp	1.292132	0.089774	3.69	0.000	1.127632	1.480629
rfq	1.862187	0.140173	8.26	0.000	1.606758	2.158221
Latvian	0.5575813	0.0691721	−4.71	0.000	0.4372304	0.7110596
Override	0.609279	0.1086291	−2.78	0.005	0.4295903	0.8641277
Top11	1.578956	0.2179024	3.31	0.001	1.204761	2.069376
preaward	1.300524	0.0589671	5.80	0.000	1.189938	1.421389
ADR	2.144721	0.1932751	8.47	0.000	1.797477	2.559047
NumProtesters	1.073864	0.0101712	7.52	0.000	1.054112	1.093985
_cons	0.3720372	0.0328634	−11.19	0.000	0.3128936	0.4423602

SOURCE: RAND analysis of GAO data.

Table A.2
Logistic Analysis for Sustained Rate at GAO

. sw logistic Sustain TaskOrder small Army AirForce Navy DLA Reconsideration rfp rfq Latvian Override Top11 preaward ADR NumProtesters InitialFiling if DoD==1&protest==1, pr(0.05)

> begin with full model

p = 0.9926 >= 0.0500 removing NumProtesters
p = 0.7995 >= 0.0500 removing rfq
p = 0.3075 >= 0.0500 removing Latvian

Logistic regression			Number of observations	=	11,459
			LR chi^2(13)	=	320.57
			Prob > chi^2	=	0.0000
Log likelihood = –1,221.3402			Pseudo R^2	=	0.1160

Sustained	Odds Ratio	Standard Error	z	P>|z|	[95% Confidence Interval]	
TaskOrder	1.477047	0.239789	2.40	0.016	1.074502	2.030399
small	0.7357623	0.0957658	–2.36	0.018	0.5700939	0.9495738
Army	0.3064527	0.0544516	–6.66	0.000	0.216331	0.4341183
AirForce	0.4979394	0.0974572	–3.56	0.000	0.3392958	0.7307595
Navy	0.6691241	0.1230296	–2.19	0.029	0.4666584	0.9594322
DLA	0.2414639	0.0767306	–4.47	0.000	0.129528	0.4501328
Reconsideration	0.0632285	0.063587	–2.75	0.006	0.0088081	0.45388
rfp	1.783608	0.2956334	3.49	0.000	1.288881	2.468234
InitialFiling	0.3369774	0.043503	–8.43	0.000	0.2616451	0.4339992
ADR	0.1646215	0.083823	–3.54	0.000	0.0606829	0.4465877
Override	2.555969	0.850831	2.82	0.005	1.33109	4.907989
Top11	3.266195	0.7317866	5.28	0.000	2.105385	5.067021
preaward	0.4966043	0.0906896	–3.83	0.000	0.3471879	0.7103238
_cons	0.0820081	0.017405	–11.78	0.000	0.0541006	0.1243116

SOURCE: RAND analysis of GAO data.

Table A.3
Logistic Analysis for Sustained Rate at COFC

. sw logistic Valid Army AirForce Navy DLA YearFiled GAO NumofIntDef DaysPending DaysRecordFiled
ProcAmount Small Appeal if A
> gency!="", pr(0.05)

begin with full model

p = 0.9735 >= 0.0500 removing GAO
p = 0.8670 >= 0.0500 removing AirForce
p = 0.7072 >= 0.0500 removing Army
p = 0.4780 >= 0.0500 removing Small
p = 0.4271 >= 0.0500 removing ProcAmount
p = 0.3968 >= 0.0500 removing NumofIntDef
p = 0.2545 >= 0.0500 removing Navy
p = 0.2238 >= 0.0500 removing DLA
p = 0.2595 >= 0.0500 removing DaysPending
p = 0.2664 >= 0.0500 removing DaysRecordFiled

Logistic regression		Number of observations	=	195
		LR chi^2(2)	=	12.18
		Prob > chi^2	=	0.0023
Log likelihood = −58.394522		Pseudo R^2	=	0.0944

Valid	Odds Ratio	Standard Error	z	P>|z|	[95% Confidence Interval]	
YearFiled	0.7946723	0.079581	−2.29	0.022	0.6530494	0.967008
Appeal	3.090733	1.605943	2.17	0.030	1.116292	8.557466
_cons	5.2e+199	1.0e+202	2.28	0.022	1.87e+28	

SOURCE: RAND analysis of COFC data.

Figure A.10
Number of Active Judges at COFC, CYs 2008–2017

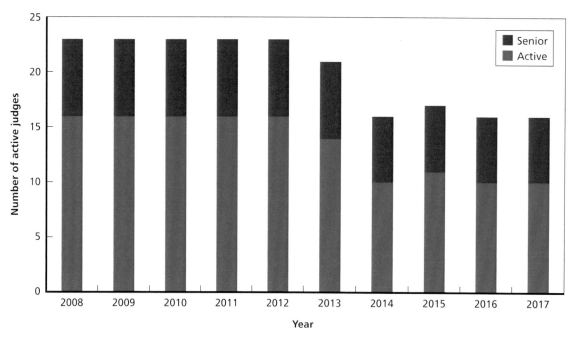

SOURCE: RAND analysis of COFC data.
RAND *RR2356-A.10*

Figure A.11
Fraction of Cases Appealed, CYs 2008–2016

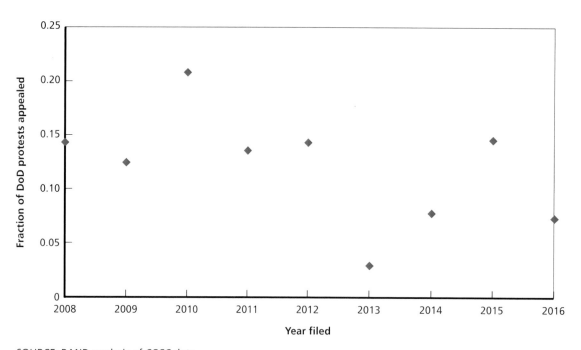

SOURCE: RAND analysis of COFC data.
RAND *RR2356-A.11*

FPDS-NG Analyses

This appendix provides background on our analyses of FPDS-NG data from FYs 2008–2016. FY 2016 was the latest complete fiscal year for which federal government contract spending data were available. FY 2008 was the first year in which the services were required to report acquisition purchases made in Iraq, Afghanistan, and Kuwait; these data are thus the earliest available that account for all DoD contract spending for overseas contingency operations (OCO).

Subsequent subsections include a description of our analytical approach, DoD spending, and DoD small business spending. We include a description of FPDS-NG and sources. Knowledgeable readers may want to skip sections on data fidelity, definitions, and approach.

Analytical Approach

Data Description, Sources, and Fidelity

The FPDS-NG website describes the statutory requirement to report all direct federal contract actions under FAR 4.6, titled "Contract Reporting," that meet certain criteria. Before FY 2016, in general, all actions related to contracts with an estimated value of more than $3,000 were reported. In FY 2016, this changed to $3,500. All actions for those contracts, even administrative actions with zero dollars, must be reported. Other dollar thresholds apply to specific categories of contracts, such as construction, but for our purposes, those thresholds are the most important.

We downloaded data from the public website USASpending.gov, which provides archived FPDS-NG data as far back as FY 2008. The website is updated continually by contracting officers, with a 90-day lag in posted information. Over the course of this research, we downloaded data several times—at the beginning of the study and near the end—to ensure that we were analyzing the most up-to-date information possible. Contract actions as old as five years can be incorporated on a continual basis as contracts finally close and records are updated or corrections are made. We updated our database near the end of the study for this reason. Our total database dollar values were within less than half a percent of the current FPDS-NG data for each fiscal year as of this writing.

Definitions and Approach

Of the many FPDS-NG data elements available, the ones that were most relevant to our analyses pertained to the Air Force, Army, Navy (the Marine Corps was considered part of the Navy), DLA, and remaining DoD-related agencies. A full list of these organizations can be found on the Defense Procurement and Acquisition Policy website.[1] They are largely civilian DoD organizations. We analyzed DLA separately because of its large number of contract actions.

FPDS-NG reports all contract actions made by DoD organizations regardless of which organization wrote the original contract. For example, the military services purchase many goods through General Services Administration (GSA) contracts, and they purchase other goods and services from each other and from other DoD organizations. The overwhelming majority of dollars, contracts, and actions derive from the organizations themselves.

We combined two variables, the Procurement Instrument Identifier and the Indefinite Delivery Vehicle, and called it the *contract number*. We used GSA-related contract numbers rather than task-order numbers whenever possible. We did the same for other contracts as well; that is, we analyzed contract numbers rather than task orders whenever possible.

We analyzed all FPDS-NG data to be as consistent as possible with the bid protest data. This meant we analyzed contract numbers, dollars, and actions by fiscal year and contract dollars in constant FY 2018 dollars. We used the variable "contracting officer business size determination" to categorize contracts as either small business–related or not small business–related. We did this because this variable was complete and the FPDS-NG data included many types of small- and minority-owned business variables that went beyond the scope of our study. For our purposes, we assumed that the contracting officer was in a good position to ascertain whether a company was a small business.

DoD Spending

We show in this section the results of our analyses of FPDS-NG total spending for FYs 2008–2016. All amounts are in FY 2018 dollars.

Looking across all DoD organizations in Figure B.1, military service dollars made up nearly 80 percent of all contract dollars over the FY 2008–FY 2016 period, with the Army making up most of the dollars of all organizations, followed by the Navy and the Air Force. The Army was the executive agent for contracting over most of this period in Iraq, Afghanistan, and Kuwait and as OCO dollars decrease, Army-related dollars will likely decrease as well.

Figure B.2 shows dollars by organization type and fiscal year. As mentioned earlier, the Army was assigned as the executive agent for contracting in U.S. Central Command during Operation Iraqi Freedom and Operation Enduring Freedom in Afghanistan, as well as the staging area in Kuwait. As such, many OCO contracts, dollars, and actions are reported as Army-related. As U.S. forces returned to their home bases, contracting dollars decreased. Total dollars by fiscal year for other DoD organizations did not vary significantly over the study period.

We also analyzed the number of contracts by DoD organization type by fiscal year, as shown in Figure B.3. The figure shows a decrease in Army and Navy contracts over time with little change in the total number for the Air Force and other DoD organizations. On the other hand, the number of DLA contracts increased significantly in FYs 2015 and 2016.

---

[1]   Office of the Secretary of Defense, Defense Procurement and Acquisition Policy, "DoDAAD Tables, Codes, and Rules: 4th Estate Agencies, Field Activities, and Unified Commands," August 20, 2015.

Figure B.1
Overall Contract Dollars by DoD Organization,
FYs 2008–2016

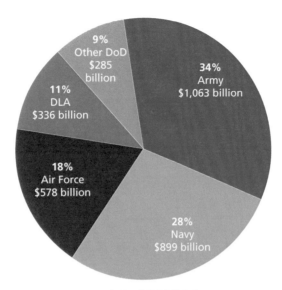

SOURCE: RAND analysis of FPDS-NG data.
NOTE: Figure shows sums in constant FY 2018 dollars.
RAND *RR2356-B.1*

Figure B.2
Contract Dollars by DoD Organization and Fiscal Year, FYs 2008–2016

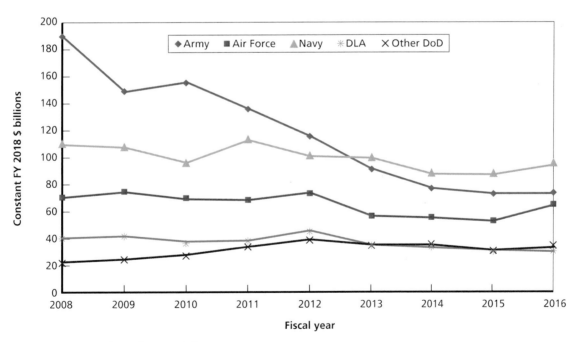

SOURCE: RAND analysis of FPDS-NG data.
RAND *RR2356-B.2*

Figure B.3
Contracts by DoD Organization, FYs 2008–2016

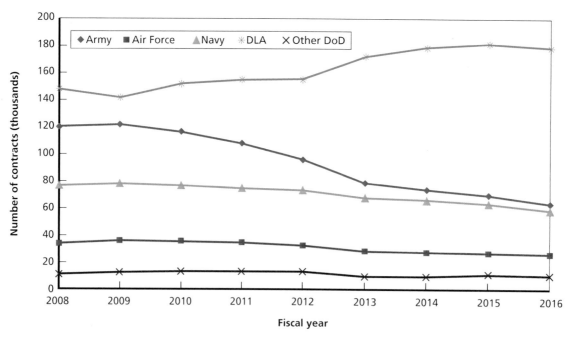

SOURCE: RAND analysis of FPDS-NG data.
RAND *RR2356-B.3*

Figure B.4
Average Number of Dollars Per Contract by DoD Organization, FYs 2008–2016

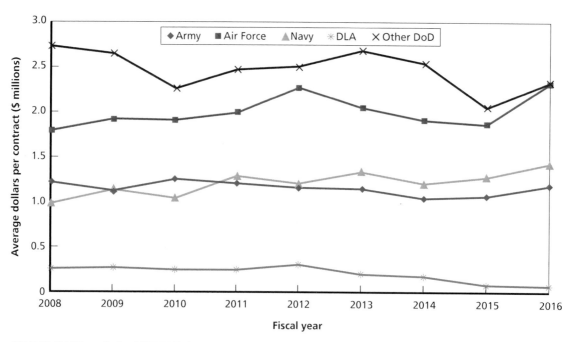

SOURCE: RAND analysis of FPDS-NG data.
RAND *RR2356-B.4*

The average number of dollars per contract by DoD organization for FYs 2008–2016, shown in Figure B.4, indicates that the average dollar value per contract was generally highest for other DoD organizations excluding DLA, followed by the Air Force, Navy, Army, and finally, DLA. High-dollar missile or health care contracts drove up the average contract dollars for other DoD organizations. Missile contract dollars for ground systems and the Missile Defense Agency are generally relatively high-value.

We also analyzed the number of contract actions for each DoD organization type by fiscal year, as shown in Figure B.5. The FPDS-NG data included a variable called "number of actions," which we summed to tally the total number of contract actions. Actions can include obligations or deobligations with associated dollars or administrative events with a cost of zero dollars that are still reported.

The total number of actions increased significantly for other DoD organizations, excluding DLA, beginning in FY 2010. DLA actions surged beginning in FY 2014. For other DoD organizations with the largest number of actions, contracts were associated with commissary and U.S. Transportation Command.

DoD Small Business Spending

We also analyzed the total number of contract dollars and the number of contracts by organization type and fiscal year and plotted them in a scatter plot in Figure B.6. Each data point represents an organization type and fiscal year. For each, we summed dollars and contracts and plotted them accordingly. The data points in the blue ovals are "other than small businesses,"

Figure B.5
Number of Contract Actions by DoD Organization Type and Fiscal Year, FYs 2008–2016

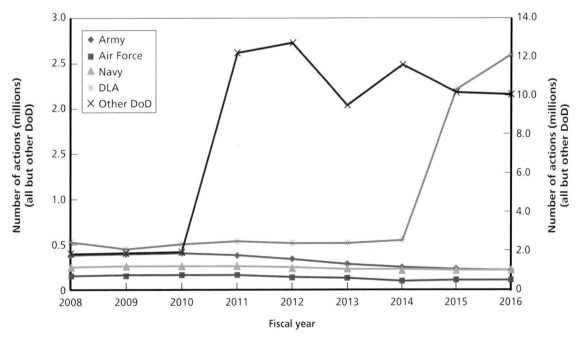

SOURCE: RAND analysis of FPDS-NG data.
RAND RR2356-B.5

Figure B.6
Dollars and Number of Contracts Across DoD Organizations Awarded to Small and Non-Small Businesses, FYs 2008–2016

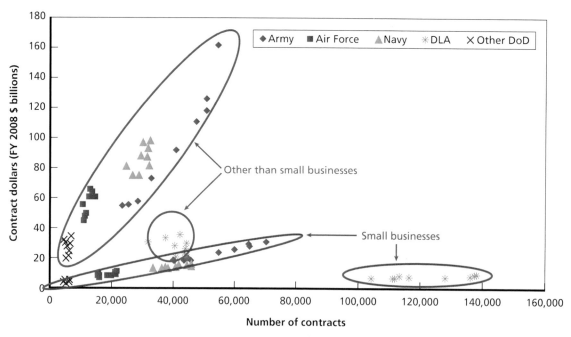

SOURCE: RAND analysis of FPDS-NG data.
RAND *RR2356-B.6*

and the data points in the red ovals are "small businesses." In general, more dollars are spent on non–small businesses. The number of contracts by fiscal year and organization can vary substantially for both sets of data points.

DoD Discussion Questions

The following questions guided our discussions with DoD personnel.

1. What is your sense of recent trends in bid protests?
   - Are protests increasing or decreasing in rate and number?
   - Is this affected by changes in procurement funding?
   - Are companies more or less likely to file a protest as a normal course of their business strategy?
   - Are incumbents more likely to protest after losing a bid?

2. How do you track bid protests?
   - What kinds of data do you keep?
   - How far back do these data go?
   - Are there reports that you can share?

3. Do you think that the specter of a bid protest influences acquisition decisions in terms of how RFPs are structured or evaluated? Does the fear of a bid protest limit acquisition/contracting options?

4. When an accommodation is made, how is this decided and documented? Are these accommodations based mainly on merits, or are they used as a way to mitigate the effect of the protest action on the program?

5. How many "bridge contracts" has your organization established in the past five years? How many of these bridge contracts were established because of a bid protest?

6. How often does your organization provide debriefings to the companies that lost a bid? Does your organization believe that the threat of a bid protest influences the quality/quantity of the debriefings? Does the quality of the debriefings have an impact on whether a bid protest is filed or not filed?

7. Do you track resources involved when there is a protest?

8. Do you track the time organizations spend taking corrective actions (for a successful protest) or overall delays to programs as a result of protests?

9. How often does the protesting firm/entity wind up winning the contract eventually?

10. Have you made any changes to acquisition practices to reduce protests in the past few years? If so, what are the changes?

11. Do you think that penalties for unsuccessful protesters would bring any improvement to the process? Explain why or why not.

12. Are there other changes to the bid protest system that would improve the overall contracting/acquisition process?

13. Is there anything we failed to ask that would assist us in our review?

References

Buettner, Douglas J., and Philip S. Anton, *Bid Protests on DoD Source Selections*, Office of the Under Secretary of Defense for Acquisition, Technology, and Logistics, Performance Assessments and Root-Cause Analyses, June 2017.

Camm, Frank, Mary E. Chenoweth, John C. Graser, Thomas Light, Mark A. Lorell, and Susan K. Woodward, *Government Accountability Office Bid Protests in Air Force Source Selections: Evidence and Options—Executive Summary*, Santa Monica, Calif.: RAND Corporation, MG-1077-AF, 2012. As of December 1, 2017: https://www.rand.org/pubs/monographs/MG1077.html

Carey, Jay, and Kevin Barnett, "GAO's Task Order Protest Jurisdiction Expires Today," *Inside Government Contracts*, September 30, 2016. As of November 19, 2017: https://www.insidegovernmentcontracts.com/2016/09/gaos-task-order-protest-jurisdiction-expires-today

Clark, Charles S., "Conferees Will Determine Fate of Defense Bill Provision to Deter Frivolous Contractor Bid Protests," *Government Executive*, October 13, 2017. As of December 12, 2017: http://www.govexec.com/contracting/2017/10/defense-bill-conferees-will-determine-fate-provision-deter-frivolous-contractor-bid-protests/141772

Coburn, George M., "The New Bid Protest Remedies Created by the Competition in Contracting Act of 1984," *Journal of Contract Management*, Vol. 19, Summer 1985.

Code of Federal Regulations, Title 48, "Federal Acquisition Regulations System," Section 33.103, "Protests to the Agency."

Davenport, Christian, "With Budget Tightening, Disputes over Federal Contracts Increase," *Washington Post*, April 7, 2014.

———, "Senate Proposes Measure to Curb Protests over Pentagon Contract Awards," *Washington Post*, October 8, 2017.

Deltek, "GovWinIQ: Grow Your Business and Bottom Line—Overview," webpage, undated. As of December 5, 2017: https://www.deltek.com/en/products/business-development/govwin

Executive Order 12979, *Agency Procurement Protests*, October 25, 1995. As of December 4, 2017: https://www.gpo.gov/fdsys/pkg/WCPD-1995-10-30/html/WCPD-1995-10-30-Pg1943.htm

FAR—*See* Federal Acquisition Regulation.

Federal Acquisition Regulation, Subpart 33.1, "Protests," May 29, 2014.

Federal Business Opportunities, "Frequently Asked Questions," webpage, undated. As of December 5, 2017: https://www.fbo.gov/?s=getstart&mode=list&tab=list&tabmode=list&static=faqs

Federal Procurement Data System: Next Generation, "Top 100 Contractors Report, Fiscal Year 2016," spreadsheet, undated. As of December 5, 2017: https://www.fpds.gov/downloads/top_requests/Top_100_Contractors_Report_Fiscal_Year_2016.xls

GAO—*See* U.S. Government Accountability Office.

Gordon, Daniel I., "Constructing a Bid Protest Process: Choices That Every Procurement Challenge System Must Make," *Public Contract Law Journal*, Vol. 35, No. 3, Spring 2006.

————, "Bid Protests: The Costs Are Real, but the Benefits Outweigh Them," *Public Contract Law Journal*, Vol. 42, No. 3, Spring 2013.

Hawkins, Timothy G., Cory Yoder, and Michael J. Gravier, "Federal Bid Protests: Is the Tail Wagging the Dog?" *Journal of Public Procurement*, Vol. 16, No. 2, Summer 2016, pp. 152–190.

Interagency Alternative Dispute Resolution Working Group, "Electronic Guide to Federal Procurement ADR 2d," webpage, undated(a). As of December 4, 2017:
https://www.adr.gov/adrguide/home.html

————, "Electronic Guide to Federal Procurement ADR 2d: Ch.08—Improved Debriefings," webpage, undated(b). As of December 4, 2017:
https://www.adr.gov/adrguide/ch08.html

Khoury, Paul F., Brian Walsh, and Gary S. Ward, "A Data-Driven Look at the GAO Protest System," *Pratt's Government Contracting Law Report*, Vol. 3, No. 3, March 2017, pp. 83–91.

Konkel, Fred, "Bid Protests Decline in 2017," *Nextgov*, November 15, 2017. As of December 4, 2017:
http://www.nextgov.com/cio-briefing/2017/11/bid-protests-decline-2017/142576

Koprince, Steven, "150 Protests and Counting: GAO Suspends 'Frequent Protester,'" *SmallGovCon*, August 22, 2016. As of December 7, 2017:
http://smallgovcon.com/gaobidprotests/150-protests-and-counting-gao-suspends-frequent-protester

————, "Senate 2018 NDAA Re-Introduces Flawed GAO Bid Protest 'Reforms,'" *SmallGovCon*, July 28, 2017. As of December 12, 2017:
http://smallgovcon.com/statutes-and-regulations/senate-2018-ndaa-re-introduces-flawed-bid-protest-reforms

Kovacic, William E., "Procurement Reform and the Choice of Forum in Bid Protest Disputes," *Administrative Law Journal of American University*, Vol. 9, 1995, pp. 461–514.

Levine, Alex, "While Government Spending Is Down, Bid Protests Are Up," blog post, PilieroMazza, September 18, 2015. As of December 7, 2017:
http://www.pilieromazza.com/blog/while-government-spending-is-down-bid-protests-are-up

Light, Thomas, Frank Camm, Mary E. Chenoweth, Peter Anthony Lewis, and Rena Rudavsky, *Analysis of Government Accountability Office Bid Protests in Air Force Source Selections over the Past Two Decades*, Santa Monica, Calif.: RAND Corporation, TR-883-AF, 2012. As of December 1, 2017:
https://www.rand.org/pubs/technical_reports/TR883.html

Manuel, Kate M., and Moshe Schwartz, *GAO Bid Protests: An Overview of Time Frames and Procedures*, Washington, D.C.: Congressional Research Service, R40228, January 19, 2016.

National Defense Industrial Association, *Pathway to Transformation: NDIA Acquisition Reform Recommendations*, Arlington, Va., November 14, 2014. As of December 3, 2017:
https://www.ndia.org/-/media/sites/ndia/policy/documents/acquisition-reform/acquisition-reform-initiative/ndia-pathway-to-transformation-acquisition-report-1.ashx?la=en

Office of the Secretary of Defense, Defense Procurement and Acquisition Policy, "DoDAAD Tables, Codes, and Rules: 4th Estate Agencies, Field Activities, and Unified Commands," August 20, 2015. As of December 11, 2017:
https://www.acq.osd.mil/dpap/sa/docs/fde/OASIS/4th_Estate_Organizations-DoDAAD_Table-20August2015.pdf

————, "Government Purchase Card (GPC) Frequently Asked Questions (FAQs)," webpage, last updated October 17, 2017. As of December 4, 2017:
https://www.acq.osd.mil/dpap/pdi/pc/faq.html

Office of the Under Secretary of Defense for Acquisition, Technology, and Logistics, *Performance of the Defense Acquisition System: 2015 Annual Report*, Washington, D.C., September 16, 2015.

————, *Performance of the Defense Acquisition System: 2016 Annual Report*, Washington, D.C., October 24, 2016.

Oliver, Richard B., and David B. Dixon, "Changes for Bid Protests in FY 2018 NDAA," Pillsbury Winthrop Shaw Pittman LLP, November 16, 2017. As of December 4, 2017:
https://www.pillsburylaw.com/en/news-and-insights/changes-bid-protest-2018-ndaa.html

Oliver, Richard B., Alexander B. Ginsberg, and Selena Brady, "Differing GAO Task Order Protest Thresholds," Pillsbury Winthrop Shaw Pittman LLP, January 3, 2017. As of December 18, 2017:
https://www.pillsburylaw.com/en/news-and-insights/differing-gao-task-order-protest-thresholds.html

Public Law 114-328, National Defense Authorization Act for Fiscal Year 2017, December 23, 2016.

Public Law 115-91, National Defense Authorization Act for Fiscal Year 2018, December 12, 2017.

Saunders, Raymond M., and Patrick Butler, "A Timely Reform: Impose Timeliness Rules for Filing Bid Protests at the Court of Federal Claims," *Public Contract Law Journal*, Vol. 39, No. 3, Spring 2010, pp. 539–581.

Schaengold, Michael J., Michael Guiffre, and Elizabeth M. Gill, "Choice of Forum for Federal Government Contract Bid Protests," *Federal Circuit Bar Journal*, Vol. 18, No. 243, 2009.

Schwartz, Moshe, and Kate M. Manuel, *GAO Bid Protests: Trends and Analysis*, Washington, D.C.: Congressional Research Service, R40227, July 21, 2015.

Serbu, Jared, "Senate Backs Down from Attempt to Restrain Bid Protests, but Wants More Data," Federal News Radio, December 5, 2016. As of December 4, 2017:
https://federalnewsradio.com/dod-reporters-notebook-jared-serbu/2016/12/senate-backs-attempt-restrain-bid-protests-wants-data

StataCorp LLC, "logit—Logistic Regression, Reporting Coefficients," undated. As of December 5, 2017:
https://www.stata.com/manuals13/rlogit.pdf

U.S. Army, *Override of CICA Stays: A Guidebook*, version 3, June 2008.

U.S. Code, Title 41, Public Contracts, Section 321, Limitation on Pleading Contract Provisions Related to Finality; Standards of Review.

U.S. Environmental Protection Agency, "Summary of the Administrative Procedure Act: 5 USC §551 et seq. (1946)," webpage, last updated December 30, 2016. As of December 5, 2017:
https://www.epa.gov/laws-regulations/summary-administrative-procedure-act

U.S. Government Accountability Office, "Bid Protest Annual Reports," webpage, undated(a). As of December 5, 2017:
https://www.gao.gov/legal/bid-protest-annual-reports/about

———, "Bid Protests at GAO: A Descriptive Guide," webpage, undated(b). As of December 19, 2017:
https://www.gao.gov/decisions/bidpro/bid/filing.html

———, "Bid Protests: Search Protests," webpage, undated(c). As of December 5, 2017:
https://www.gao.gov/legal/bid-protests/search

———, "Bid Protests: Our Process," webpage, undated(d). As of December 5, 2017:
https://www.gao.gov/legal/bid-protests/our-process

———, *Bid Protests at GAO: A Descriptive Guide*, 9th ed., Washington, D.C., GAO-09-471SP, 2009a. As of December 7, 2017:
https://www.gao.gov/assets/210/203631.pdf

———, *Report to Congress on Bid Protests Involving Defense Procurements*, Washington, D.C., B-401197, April 9, 2009b. As of December 7, 2017:
http://www.gao.gov/decisions/bidpro/401197.pdf

Yang, David, "Senate Proposes Major Overhaul to the GAO Bid Protest Process," *Government Contracts Navigator*, October 3, 2017. As of December 7, 2017:
https://governmentcontractsnavigator.com/2017/10/03/senate-proposes-major-overhaul-to-the-gao-bid-protest-process